BURLEIGH DODDS SCIENCE: INSTANT INSIGHTS

NUMBER 70

Economics of key agricultural practices

burleigh dodds
SCIENCE PUBLISHING

Published by Burleigh Dodds Science Publishing Limited
82 High Street, Sawston, Cambridge CB22 3HJ, UK
www.bdspublishing.com

Burleigh Dodds Science Publishing, 1518 Walnut Street, Suite 900, Philadelphia, PA 19102-3406, USA

First published 2023 by Burleigh Dodds Science Publishing Limited

British Library Cataloguing in Publication Data
A catalogue record for this book is available from the British Library

ISBN 978-1-80146-602-8 (Print)
ISBN 978-1-80146-603-5 (ePub)

DOI: 10.19103/9781801466035

Typeset by Deanta Global Publishing Services, Dublin, Ireland

Contents

Series list

Title	Series number
Sweetpotato	01
Fusarium in cereals	02
Vertical farming in horticulture	03
Nutraceuticals in fruit and vegetables	04
Climate change, insect pests and invasive species	05
Metabolic disorders in dairy cattle	06
Mastitis in dairy cattle	07
Heat stress in dairy cattle	08
African swine fever	09
Pesticide residues in agriculture	10
Fruit losses and waste	11
Improving crop nutrient use efficiency	12
Antibiotics in poultry production	13
Bone health in poultry	14
Feather-pecking in poultry	15
Environmental impact of livestock production	16
Pre- and probiotics in pig nutrition	17
Improving piglet welfare	18
Crop biofortification	19
Crop rotations	20
Cover crops	21
Plant growth-promoting rhizobacteria	22
Arbuscular mycorrhizal fungi	23
Nematode pests in agriculture	24
Drought-resistant crops	25
Advances in detecting and forecasting crop pests and diseases	26
Mycotoxin detection and control	27
Mite pests in agriculture	28
Supporting cereal production in sub-Saharan Africa	29
Lameness in dairy cattle	30
Infertility/reproductive disorders in dairy cattle	31
Alternatives to antibiotics in pig production	32
Integrated crop-livestock systems	33
Genetic modification of crops	34

Chapter 1

Assessing the economics of integrated pest management for horticultural crops

Philip R. Crain and David W. Onstad, Corteva Agriscience, USA

1 Introduction

Assessing the value and strategic implications of integrated pest management (IPM) is a challenging but necessary exercise. Economic value, which can be defined as an operating loss or gain on a crop in a given period of time, is a very real observation (i.e. one point of data) that carries significant consequences for farmers. Economic value, however, is a multifaceted, variable, and dynamic concept. For example, the loss or profit in a given year at a given location is subject to many forces, some sustained and some ephemeral. The challenge to IPM is to help optimize economic value while designing robust pest management strategies for farmers and other stakeholders, including those concerned about environmental impacts. These strategies should address known risks and be flexible enough to quickly adapt to emerging risks.

Optimized economic value is obtained when farmers can meet their short-term growth targets that sustain and improve their livelihood while also addressing the need to produce more in environmentally and financially sustainable ways. Concepts in IPM, like economic injury level and economic threshold, are conceptual models that seek to optimize economic value in the face of pests. By only using control tactics when the economic threshold had been reached, in theory, growers and control decision makers were being economical in their decisions. However, what if the control tactic was

http://dx.doi.org/10.19103/AS.2021.0095.12

more expensive than a change in the design of the IPM system? What if a new technology, for example exclusion nets, provided a more reliable, ecologically sustainable, and more valuable control compared to the overuse of a chemical insecticide? This highlights a significant dilemma facing growers and pest control advisors, which is: when should we 'think inside the box' of current pest management paradigms, and when should we explore 'outside the box' to unlock better value or sustainability?

'Thinking inside the box' tends to have a negative connotation. It is somehow viewed as less inspired than those solutions that 'think outside the box.' There have been many great successes that have resulted from thinking inside the box. 'The box' in this metaphor represents a system or framework that is useful in describing the attributes of a system and the most common decisions that are confronted. New control tactics, which may arise from inside the box thinking, address a known need to manage risk and create value for stakeholders. Both current and new control tactics are less likely to create value when new risks arise. The emergence of unknown risks may have been driven by the selection pressure exerted by control tactics or it could represent new pest invasions. To manage new risks, stakeholders need to think differently, consider disruptive technologies, and explore system design changes to find new solutions that maintain economic value. Economic thinking is critical to making good decisions that not only balance potential benefits but also realize the potential costs of decisions (i.e. optimizing economic value).

Economic thinking that focuses on the return on investment of control strategies in agricultural systems and specifically IPM is difficult, however. The complex and chaotic nature of ecological interactions lends itself to rule of thumb concepts that simplify the complexity. Simplifying complex systems can be too easily interpreted as a disruptive innovation. It distills difficult *n*-dimensional systems into a set of manageable, useful choices. This provides benefit to practitioners because it allows them to measure specific attributes of the system that can guide decisions that are likely better. The risk is that these systems become static. Because agricultural systems are dynamic, static frameworks are useful only for a specific window of time. IPM practitioners must consciously think about the complexity and dynamic nature of systems to build conceptual models that distill the most important system attributes into a series of decisions that can positively impact the outcome of farmers and society.

Despite the dynamic and complex nature of agricultural problems, the basic components of economic thinking are extremely helpful to promote careful consideration of pest management strategies. In its most basic sense, economic thinking is confronting the assumptions that stakeholders hold about the benefit (i.e. revenue) and costs of decisions they take within their system. The number of unknowns at the start of a season is very high, and each decision a farmer takes should address an unknown. Each decision has a potential positive and negative impact. It is critical to weigh these concepts to make good pest management decisions. Of course, the set of unknowns that exist before a single season, and which focus on a single season, is different from the set of unknowns that dictate more long-term questions like resistance management (Onstad 2013). Economic thinking can also be used to understand how to focus on short- and long-term goals to maintain viability and sustainability of operations.

In this chapter, we introduce the concepts of economic thinking, describing the costs and potential benefits of IPM control solutions as well as discussing time horizons, stakeholders, and economic injury level and economic thresholds. Next, we will look at some case studies, with a focus on horticulture, which exemplify how to measure ecological complexity through economic thinking. Throughout the discussion of economic principles and case studies, we will highlight opportunities to consider control tactics versus IPM system design changes and the implications for economic decision-making. Finally, we will address the future of economic thinking in IPM with a challenge that creates more value for society and growers in the future through innovation and new technology that is becoming available.

2 Concepts of economic thinking

Economics is broadly the study of human choice under the limits of scarce resources (Sowell 2014). More specifically, economics is often defined as the branch of knowledge that focuses on the production, consumption, and transfer of wealth. While economics does deal with value, wealth, and currency, it is not limited to viewing components of a system as either costs or benefits financially. Economics also must consider human values and how they are reflected in decision-making. This generally describes normative economics, or the analysis of what 'ought to be' (one can also focus on positive economics or 'what is'). The goods and services that people use carry with them assumptions for how they are valued and how they are perceived. For example, entomologists may recognize that an arthropod pest is a subjective, human-centric definition that only applies to specific places at specific times. Insects can be beautiful and indicative of a healthy ecosystem in some contexts but hurtful in another. For example, European honeybees (*Apis mellifera*) are generally regarded as beneficial insects due to the pollination services they provide, but they may be considered pests if a hive is too close to humans or if they are disrupting the natural bees present in an ecosystem.

To perform an economic analysis, it is important at the outset to rigorously define the problem. Components of a problem are (1) identifying the stakeholders involved and what is the primary perspective being taken, (2) aligning the goals of the analysis with the perspective of the primary stakeholder(s), and (3) defining the time period and spatial scale for the analysis as much as possible to expose any risks or dependencies. As mentioned previously, the decisions, benefits, and costs that characterize a single year or crop-growing season may be very different from those associated with a multiyear or long-term outlook. Economic thinking helps to align stakeholders' expectations, costs, benefits, and time horizons so that decision makers can use this information to allocate resources and manage pests to achieve multiple objectives in the lives and well-being of people and economies (National Research Council 1999, 2005)

To illustrate the scope of this challenge, consider two different examples. First, vineyard owners have an established vineyard and are looking to maximize their return for the upcoming year. Because vineyards are perennial crops, there is no decision to take on what crop should be planted, but instead, their focus will look to predicting pest incidence and what tools they can use to control those problems before they cause economic damage. Along with those decisions, they must consider how they will market their product and if that constrains their possible decisions. For example, organic production will have a very different toolkit for controlling pests relative to conventional vineyards. Grapes for wine production may have different considerations versus table grapes. Many decisions may be taken before the season begins, but many decisions will also occur within the season based on the weather and conditions experienced by their crop that year. To create profit, the owners must generate enough revenue to cover the costs of any control treatments, labor, equipment or land costs, and many other factors. A second stakeholder could be a consumer of the products from this vineyard. Being focused on environmental impacts and long-term sustainability, consumers could create secondary considerations for the vineyard grower in the long-term. Maybe they need to shift the acres in their production to more organic to accommodate consumer demand? Maybe grape demand is falling, and they will need to confront the costs associated with picking a new crop? These are different questions arising from various stakeholders with different time horizons. Defining problems within these spaces is critical to solving the right problem with economic thinking.

In IPM, stakeholders range from growers to consultants or advisors to input and processing companies to product consumers. All these individuals are influenced by IPM decisions. Growers typically bear the costs of any control tactic used, while processing companies and consumers may define market opportunities and their value (e.g. premium prices for organic production vs. conventional). As new market opportunities become available, growers must carefully consider the return they may generate for the cost they will incur and the time horizon for both. For example, while organic production is more desirable for many consumers and may lead to higher prices for produce, the cost of organic production is generally higher with the need to use more control tactics in combination coupled with less efficacy for each individual tactic. Akin and Kara (2021) found that willingness to continue organic almond production in Turkey decreased when orchard size increased because the value of additional production in an additional hectare of land was more valuable to growers than the payments they received to maintain organic production (organic production has lower yields relative to conventional production in this system). Walsh et al. (2011) performed an economic assessment of an experimental orchard with apple and pear trees that included both organic and conventional growing practices. Their analysis found that organic production required more labor and inputs relative to conventional production and estimated yield was also lower in organic blocks. Understanding the relationships among stakeholders and the focal perspective they hold allow us to solve problems, but stakeholders and their perspectives may change (e.g. a grower may consider different cropping systems if they add more hectares to their farm) as markets evolve.

Time horizon is the length of time over which a problem will be assessed. The decision to focus on a single year, multiple years, or a generation can have important ramifications on the decisions made, relative to their costs and benefits. Longer time horizons are not only associated with increased uncertainty but also provide more time to potentially reap benefits from control tactics or system design changes that are more costly upfront but provide ample returns after initiation. However, short-term profit generation is key to providing additional resources that growers can use to increase opportunities for future investments that may have more far-reaching outcomes. For example, purchasing additional land can help increase revenue but requires some free cash for deposits as well as financing loan payments. It is important to recognize that many growers are under significant short-term pressure to limit costs while maintaining or increasing their profitability but that social and environmentally sustainable solutions have long-term cycles that need to also be addressed in viable ways.

Within the context of horticultural cropping systems, and agriculture more generally, readers may be aware of the concepts of economic injury levels (EIL) and economic thresholds (ET). These concepts have been implemented in many systems and represent a simple interpretation of agriculture through an economic perspective. EIL is the lowest population density that will cause economic damage, while the ET is the population density where control measures should be applied to stop a pest from reaching or exceeding the EIL (Stern et al. 1959). These definitions allow for empirical studies to quantify both EIL and ET which, in theory, can lead to effective, economic control. This system is often focused on single-season dynamics and control versus system design opportunities and multiyear cropping systems. It is important to maintain the long-term view as well, especially if the crop of interest is a perennial crop where design changes may be more tractable, given the upfront costs and long-term characteristics of a particular cropping system (Hoyt and Gilpatrick 1976, Westigard 1979).

Complementary to the ET and EIL concepts, economic thinking should also allow for the consideration of system design changes. System design considers the whole agroecosystem and challenges whether the structures and/or components of a system (e.g. land, crops, cropping systems, etc.) could be changed to achieve desired outcomes more economically (Onstad and Crain 2019). Barzman et al. (2015) introduced the eight principles of pest management and stressed the importance to design cropping systems such that they only use harsh crop protection products when they best fit the needs of a grower. The first principle focuses on prevention and suppression and discusses how crop choice, management practices, and structural layout can suppress many pest problems before they become significant issues. Sciarretta et al. (2011) evaluated the effect of site-specific IPM to manage *Lobesia botrana* in grapes in central Italy and concluded that identifying and managing spatial hotspots can be more cost-effective than uniform management practices and that monitoring border olive groves helped inform grape management. Understanding the agroecosystem and pest dynamics in both time and space should allow stakeholders to identify when system design changes are economical versus their current practices.

3 Economic impact of horticultural crops and case studies on the complexity of integrated pest management

3.1 Overview of horticultural economics

Armed with the concepts of economic thinking, we will review some general economics for horticulture as well as use several case studies to dive more deeply into the themes of design and control, stakeholder engagement, and time horizons. The following sections will highlight the scope and value of horticulture in a few geographies and then transition to several case studies that can serve as examples for the themes above.

Fruit and vegetable production in Europe represented nearly 14% of the total value of agricultural production in 2018 or over €57B (Rossi 2019). Fruits accounted for 40% of the total value and vegetables the remaining 60%. Of total production tonnes, tomatoes and apples are the most significant contributors, and each represents approximately 20% of the total value of vegetables and fruits. The costs associated with pest management to produce those quantities are difficult to estimate because many control options are not easily quantifiable or tracked by government agencies or industry partners (e.g. agricultural practices that contribute to cultural control). However, Wesseler et al. (2015) measured the input costs associated with plant protection agents, which largely reflected the costs of chemical crop protection products, and found that plant protection agents accounted for more than €7B in 2010 in Europe.

In the United States, vegetable production totaled more than $19B in 2019 (Lucier and Davis 2020). Fruit, tree nut, and berry production totaled more than $28B in 2017. Again, it was difficult to quantify the costs of inputs, but the National Road Map for Integrated Pest Management (FIPMCC 2018) suggested that cost-benefit analyses be holistic and not limited to strictly financial metrics.

A review of agriculture in Brazil by Oliveira et al. (2014) predicted total crop losses as well as spending on crop protection products to protect yields against insects across broadacre row crops and specialty crops. Sugarcane had the highest economic loss (combining use as sugar and ethanol, ~$6B), followed by other crops like corn (~$2B) and soybean (~$2B). The total loss for these crops is proportional to the large acreage they cover. If costs are evaluated per hectare, then apples ($4281/ha), tomatoes ($3807/ha), tobacco ($2730/ha), and garlic ($2655/ha) are most susceptible to yield loss from insects. In total, insecticide sales alone totals nearly $3B with a significant portion of that spent focusing on soybeans (~$1.3B).

3.2 Case studies

The invasion of *Drosophila suzukii*, the spotted wing Drosophila (SWD), has led to many efforts to build economic IPM plans to control the pest and limit or slow its geographical spread (Asplen et al. 2015). SWD prefer ripening fruit and laying eggs under the fruit's skin, resulting in unmarketable fruit (Atallah et al. 2014). A study by Mazzi et al. (2017) discussed different IPM tactics for controlling SWD and documented their potential costs in sweet cherries in Switzerland. They surveyed growers in 2015 and 2016 and learned that the most widely used tactics to manage damage from SWD were (1) insecticide applications and (2) harvest practices aimed at collecting the majority of fruits, reducing any unharvested fruits that may harbor SWD. They also mentioned exclusion nets were prevalent in a significant proportion of survey respondents (approximately 1/3). The costs of the different methods were significant, and they called for increased stakeholder engagement to focus on both the value created by control measures and the science needed to technically vet the solutions.

A second study by Santoiemma et al. (2020) also examined which control tactics best reduce *D. suzukii* infestations in sweet cherry orchards in Italy. Similar to Mazzi et al. (2017), Santoiemma et al. (2020) found that several orchard management practices provide good protection against SWD damage; specifically, the high frequency of mowing interrow vegetation and good harvest practices, covering both the timing of harvest and completeness of removing all fruit from the field, correlated with less damaged fruit. They did not see a reduction in fruit damage based on the use of insecticides.

Both the previous studies outline the challenge of managing SWD, the tactics that seem to work best, and even the costs to implement those tactics. However, there also needs to be careful consideration of the benefits to growers (and other stakeholders) for implementing these tactics. An economic analysis by Del Fava et al. (2017) used partial budget analysis in a cost–benefit analysis to investigate the expected net present value of different IPM systems following the invasion of SWD into soft fruit crops in Northern Italy in 2009. Initially, a conventional IPM strategy was employed while fly population densities were low. However, as populations continued to increase, the value of current control tactics was insufficient. Growers opted to introduce exclusion nets, despite significant implementation cost, and Del Fava et al. (2017) showed that exclusion nets resulted in increases of profit up to €2.5 million.

A similar series of studies on the area-wide pest management effort to control Queensland fruit fly, *Bactrocera tryoni,* in Australia, provides an opportunity to think more about the benefits of different control measures and how those benefits influence resource allocation and opportunities for success. The Queensland fruit fly is a significant economic pest of horticultural goods (Dominiak 2019). *B. tryoni* is endemic to Australia but is a key target for exports due to the risk of its invasion and establishment in other areas, leading to the development of a code of practice for the management of Queensland fruit fly (Dominiak et al. 2015, Standing Committee on Agriculture and Resource Management 1996). Areas of Australia (Sunraysia, or northwestern Victoria and southwestern New South Wales) have been designated as pest-free areas and exclusion zones, and these areas are granted more access to export and domestic markets relative to other areas where the pest is endemic. An economic analysis by Florec et al. (2013) evaluated the economics of area-wide pest management against this pest focusing on the eradication of Queensland fruit fly outbreaks as well as post-harvest treatments. The analysis found that reducing the probability of outbreaks by enforcing stricter control at roadblock sites, thus reducing post-harvest disinfection insecticide treatments, was more economical compared to better surveillance or faster eradication. The authors concluded that improving the ability to identify, quantify, and separate produce needing these treatments by geographical area is important to effectively managing the budget for management. Although economic benefits are difficult to quantify and require aligned assumptions among stakeholders, decisions supported by science and expected benefits will allow for allocation of resources in areas that are most likely to impact growers and society.

In cases where pest management is particularly challenging, it is worthwhile to consider innovative control techniques or system design changes that may provide unique benefits. Nikolouli et al. (2020) investigated the potential for sterile and incompatible insect techniques to aid in the suppression of SWD. Using irradiation and *Wolbachia* sp., the authors documented high levels of sterilization or incompatibility without significantly affecting the life history or behavior of sterile males. The results suggest these techniques may contribute to an IPM program, but the economic costs and benefits need to be further quantified based on limited field trials. Although not an example from horticulture, Alphey et al. (2011) found that the cost-effectiveness for using sterile insect technique coupled with the genetic control technology RIDL (release of insects carrying a dominant lethal) was much higher than conventional vector-borne disease-control strategies. Novel changes to the design of a system can dramatically improve economic pest management. However, during the early technical validation of new or innovative IPM practices, there should be funding to support the testing and scaling of technical solutions at grower operations to bring better alignment with technical solutions and practical operations.

Agriculture represents an ecological system that can be manipulated by humans, but it is also subject to the uncertainty of weather, evolution, and species interactions. Additionally, agriculture is governed by consumer demands, market trends, and the need to increase sustainability. A study by Rodriguez Sousa et al. (2020) examined the economics of ecological farming in olive production in southwestern Spain and compared it to a more conventional farming practice. They used an analytic hierarchy process to assess both the financial returns and the environmental and societal benefits of both cropping systems. The study reported that integrated (or more conventional) olive farms saw higher financial returns but that ecological-farmed olives brought higher environmental and societal benefits. In the aggregate analysis, the externalities were considered more valuable than the difference in on-farm income (~$100/ha advantage to integrated farms). Importantly, environmental and societal benefits are enjoyed by many people, but the costs are transferred to growers. We should not only continue to support transdisciplinary analyses but also find ways to drive economic well-being for the growers. Can ecological farms also serve as agroecotourism opportunities? Are farmers willing to incur the costs to make such an opportunity a reality? We advocate for economic analyses and broad stakeholder engagement to drive collaboration and better decision-making for the near and long-term.

Another consideration for economic analysis is cost of sampling to make a decision. It is not enough to consider only the costs and benefits of a treatment, but it must holistically include the data and labor needed to support a decision. A study by Figueiredo et al. (2021) took an interesting approach to improving the sampling of *Helicoverpa armigera* in tomatoes to determine if an insect control treatment is needed. They developed a sequential sampling procedure that had several criteria that either stopped sampling sooner or continued sampling to obtain acceptable precision for when to treat infestations in processing tomatoes. For a given economic threshold, the standard approach required sampling 60 plants to inform a treatment decision. By sequentially sampling the field, using the knowledge of the number of plants already sampled as well as the number of eggs, and mathematical calculations (Iwao's method based on Taylor's power law regression), scouts could stop sampling a field much sooner, sampling as few as 20 plants. By providing criteria to stop sampling and make decisions sooner, Figueiredo et al. (2021) allow scouts to be more efficient with their time. However, the cost of sampling is often ignored (Onstad et al. 2019). Building a more flexible framework that allows for more efficient decisions is an important step. Another advancement would be the coupling of sensing technology or prediction methods to further create efficiency for the grower or pest-management decision maker.

4 Conclusion

Optimizing economic value is elusive. It requires intentional focus on behalf of all stakeholders, clear and frequent communication, data collection and analysis, and an understanding of who bears costs and reaps benefits. Just as natural selection creates genotypes with desirable phenotypes, the pressures of economic viability will select for approaches that optimize value. We can expedite that optimization by employing economic thinking.

5 Future trends in research

The rise of internet of things systems, cellular connectivity, and machine learning has changed the way many people think about sensing, monitoring, and analyzing data. In earnest, IPM is the simplification of a complex system to a set of decisions that take data as input and generate sustainable revenue as an output. It is exciting to consider the impact of innovation on horticulture and agriculture more broadly, but as we have mentioned earlier, innovation comes with a cost and requires a solution that solves a problem with potentially different time horizons and stakeholders. We must help innovation in IPM succeed by making it applicable to all stakeholders in both the short and the long term.

Innovation is challenging for stakeholders to consider because disruptive innovation challenges the standard cost–benefit paradigm of cropping systems by often creating new ways of capturing value. A study by Djian-Caporalino et al. (2019) focused on evaluating several different cropping systems to understand the impact to sustainable management of root-knot nematodes in vegetable systems. By combining cultural practices, including planting nematicidal cover crops, with resistant varieties of vegetables, they found several solutions that were amenable to growers. Djian-Caporalino et al. (2019) established that more work was needed to gain efficiency for farmers, suggesting that future work should focus on innovative application of science. Using science to understand agroecological systems is critical, but understanding the costs that translate science in practical tools or methods is challenging. Economic analyses as we discussed in the previous section can help identify the areas where we can allocate resources that not only provide the best opportunity to create significant innovation but also generate profitability. Understanding the technical characteristics and attributes of new technologies as well as market opportunity (i.e. economic viability) requires intentional and dedicated thought.

A review by Fernández et al. (2018) highlights the shifts in greenhouse-based horticulture and how many of those factors are addressing emerging trends in technological and agricultural innovation. Greenhouse-based cropping systems provide a unique case to understand the benefits and costs of IPM in a more controlled environment. While the article focuses on physical structures, crops, water, and fertilization, it also discusses the need to continue innovating in IPM. The authors highlight innovations ranging from exclusion netting to biological control to a modern take on the traditional push–pull systems of IPM (see Harrison et al. (2019) for a background on push–pull systems and discussion of some agroecological management options more broadly than controlled environments). These innovations help agriculture accomplish the goal of becoming more sustainable, which is a critical task, given the importance of climate change, but they must also be profitable to the growers who must bear the cost of new operating models. Economic analyses to help understand the risks surrounding the establishment of innovative and sustainable technology, especially in the first few years, are critical. Early investment is needed to validate the science and applicability of innovative solutions and address the path from science to practical tools and methods. This early period of innovation and discovery may not result in near-term, economically viable changes for farmers and should be supported by stakeholders who would benefit from long-term adoption.

As technology and innovation is used more widely in agriculture, it will represent the opportunity to re-evaluate economic principles in IPM. As Rodriguez Sousa et al. (2020) demonstrated, value is strictly defined as a financial metric but is more encompassing to cover sustainability and quality of life. Similarly, Dara (2019) has suggested a new IPM paradigm that considers not only economic viability but also environmental safety and social acceptability. As we strive to be more sustainable, both in terms of social awareness and environment, technology and innovation offer the opportunity to collect data in new ways and apply analytics to solve complex problems. Financial viability should underlie solutions because farmers and growers must take on unique risks, so technology and innovation need to create financial benefit to support the move to broader economic criteria, including externalities.

There are many opportunities to change the design of systems, engage with stakeholders early, and define agricultural problems well to facilitate innovative solutions. The founders of IPM were aware of the challenges posed by simplifying complex systems and created important frameworks that helped turn data into action that was economical. Their initial definitions allowed for significant advancement with respect to any single control tactic within the IPM design. With the introduction of new data sources, the framework (design) of IPM needs to evolve to utilize the value of new data streams. However, the new data streams are also complex. Egea et al. (2020) discussed bioeconomy and circular economies that are driving change to greenhouse horticulture in Spain. They not only highlighted that much work has been done to create a space for sustainable and economic transformation but also signaled that there are significant needs to understand the science behind implementing bioeconomies. This work must be transdisciplinary incorporating empirical horticulture scientists, economists, data scientists and engineers, and many others. The next challenge for the stakeholders in horticulture is to take the new data streams and machine learning analytics to simplify farming operations or improve efficiency in a tractable way. For growers, the most tractable solutions reduce management complexity and increase the profit that allows them to better meet their future needs as well as society's.

6 Where to look for further information

There is a significant opportunity for individuals who wish to pursue and apply an economic perspective to IPM. There is a wide range of literature that can be used to build a good base from which many research directions can flourish.

A work that supports many of the ideas introduced here was written by the authors and is titled *The Economics of Integrated Pest Management of Insects*. Although the book focuses on insects, it broadly considers the elements of economic decision-making such as time horizon, stakeholder perspectives, and design versus control.

Agricultural economics is a discipline that will provide a rich area of exploration for those wishing to learn more. Of note, however, not all agricultural economists will consider the nuances of the economics of pest control. Notably, in the book edited by David Onstad titled *Insect Resistance Management*, Onstad and Paul Mitchell (University of Wisconsin-Madison, USA) wrote on the value of susceptibility of populations. Additional work by George Frisvold (University of Arizona, USA) and Terry Hurley (University of Minnesota) has looked at the applied economics of resistance management and integrated pest management, specifically area-wide pest management. Zack Brown (North Carolina State University, USA) is another researcher who has examined the economics of IPM and IRM.

7 References

Akin, S. and Kara, A. (2021). Factors effective on reaching goals and target groups of agricultural supports: a case study in Diyarbakir Province, Turkey. *Ciência Rural* 51(1). https://doi.org/10.1590/0103-8478cr20200260.

Alphey, N., Alphey, L. and Bonsall, M. B. (2011). A model framework to estimate impact and cost of genetics-based sterile insect methods for dengue vector control. *PLoS ONE* 6(10), e25384.

Anon. (2018). A national roadmap for integrated pest management. United States Department of Agriculture. Available at: https://www.usda.gov/media/press-releases/2018/10/24/usda-announces-update-national-road-map-integrated-pest-management (accessed 16 April 2021).

Asplen, M. K., Anfora, G., Biondi, A., Choi, D. S., Chu, D., Daane, K. M., Gibert, P., Gutierrez, A. P., Hoelmer, K. A., Hutchison, W. D., Isaacs, R., Jiang, Z., Kárpáti, Z., Kimura, M. T., Pascual, M., Philips, C. R., Plantamp, C., Ponti, L., Vétek, G., Vogt, H., Walton, V. M., Yu, Y., Zappalà, L. and Desneux, N. (2015). Invasion biology of spotted wing Drosophila (Drosophila suzukii): a global perspective and future priorities. *Journal of Pest Science* 88(3), 469–494.

Atallah, J., Teixeira, L., Salazar, R., Zaragoza, G. and Kopp, A. (2014). The making of a pest: the evolution of a fruit-penetrating ovipositor in Drosophila suzukii and related species. *Proceedings. Biological Sciences* 281(1781), 20132840.

Barzman, M., Bàrberi, P., Birch, A. N. E., Boonekamp, P., Dachbrodt-Saaydeh, S., Graf, B., Hommel, B., Jensen, J. E., Kiss, J., Kudsk, P., Lamichhane, J. R., Messéan, A., Moonen, A., Ratnadass, A., Ricci, P., Sarah, J. and Sattin, M. (2015). Eight principles of integrated pest management. *Agronomy for Sustainable Development* 35(4), 1199–1215.

Dara, S. K. (2019). The new integrated pest management paradigm for the modern age. *Journal of Integrated Pest Management* 10(1), 12.

Del Fava, E., Ioriatti, C. and Melegaro, A. (2017). Cost-benefit analysis of controlling the spotted wing drosophila (Drosophila suzukii (Matsumura)) spread and infestation of soft fruits in Trentino, Northern Italy. *Pest Management Science* 73(11), 2318-2327.

Djian-Caporalino, C., Navarrete, M., Fazari, A., Baily-Bechet, M., Marteu, N., Dufils, A., Tchamitchian, M., Lefevre, A., Pares, L., Mateille, T., Tavoillot, J., Palloix, A., Sage-Palloix, A.-M., Vedie, H., Goillon, C., and Castagnone-Sereno, P., and Castagnone-Sereno, P. (2019). Design and assessment of innovative Mediterranean vegetable cropping systems to manage root-knot nematodes description of the subject. *Biotechnologie, Agronomie, Société et Environnement* 23(1), 7–21.

Dominiak, B. C. (2019). Components of a systems approach for the management of Queensland fruit fly Bactrocera tryoni (Froggatt) in a post dimethoate fenthion era. *Crop Protection* 116, 56–67.

Dominiak, B. C., Wiseman, B., Anderson, C., Walsh, B., McMahon, M. and Duthie, R. (2015). Definition of and management strategies for areas of low pest prevalence for Queensland fruit fly Bactrocera tryoni Froggatt. *Crop Protection* 72, 41–46.

Egea, F. J., López-Rodriguez, M. D., Ona-Burgos, P., Castro, A. J., and Glass, R. (2020). Bioeconomy as a transforming driver in intensive greenhouse horticulture in SE Spain. *New biotechnology*, 61, 50–56. https://doi.org/10.1016/j.nbt.2020.11.010.

Federal Integrated Pest Management (IPM) Coordinating Committee (FIPMCC). (2018). A National Roadmap for Integrated Pest Management. Revised September 2018. PDF available at: https://ars.usda.gov/ARSUserFiles/OPMP/IPMRoad-Map-FINAL.pdf.

Fernández, J. A., Orsini, F., Baeza, E., Oztekin, G. B., Muñoz, P., Contreras, J. and Montero, J. I. (2018). Current trends in protected cultivation in mediterranean climates. *European Journal of Horticultural Science* 83(5), 294–305.

Figueiredo, E., Gonçalves, C., Duarte, S., Godinho, M. C., Mexia, A. and Torres, L. (2021). Risk assessment for tomato fruitworm in processing tomato crop-egg location and sequential sampling. *Insects* 12(1), 13.

Florec, V., Sadler, R. J., White, B. and Dominiak, B. C. (2013). Choosing the battles: the economics of area wide pest management for Queensland fruit fly. *Food Policy* 38, 203–213.

Harrison, R. D., Thierfelder, C., Baudron, F., Chinwada, P., Midega, C., Schaffner, U. and van den Berg, J. (2019). Agro-ecological options for fall armyworm (Spodoptera frugiperda JE Smith) management: providing low-cost, smallholder friendly solutions to an invasive pest. *Journal of Environmental Management* 243, 318–330.

Hoyt, S. C. and Gilpatrick, J. D. (1976). Pest management on deciduous fruits: multidisciplinary aspects. In: J. Lawrence Apple and R. F. Smith (Eds.), *Integrated Pest Management* (pp. 133–147). Springer, Boston, MA.

Lucier, G. and Davis, W. (2020). Vegetables and pulses outlook. United States Department of Agriculture. Available at: https://www.ers.usda.gov/topics/crops/vegetables-pulses/market-outlook/#:~:text=Vegetables%20and%20Pulses%20Outlook%3A%20December%202020&text=Total%20domestically%2Dgrown%20vegetable%20and,dry%20pea%2C%20and%20lentil%20crops (accessed 16 April 2021).

Mazzi, D., Bravin, E., Meraner, M., Finger, R. and Kuske, S. (2017). Economic impact of the introduction and establishment of Drosophila suzukii on sweet cherry production in Switzerland. *Insects* 8(1), 18.

National Research Council (1999). *Perspectives on Biodiversity: Valuing Its Role in an Everchanging World*. Washington, DC: The National Academies Pressd.

National Research Council (2005). *Valuing Ecosystem Services: Toward Better Environmental Decision-Making*. National Academies Press.

Nikolouli, K., Sassù, F., Mouton, L., Stauffer, C. and Bourtzis, K. (2020). Combining sterile and incompatible insect techniques for the population suppression of Drosophila suzukii. *Journal of Pest Science* 93(2), 647–661.

Oliveira, C. M., Auad, A. M., Mendes, S. M. and Frizzas, M. R. (2014). Crop losses and the economic impact of insect pests on Brazilian agriculture. *Crop Protection* 56, 50–54.

Onstad, D. W. (Ed.) (2013). *Insect Resistance Management: Biology, Economics, and Prediction*. London: Academic Press.

Onstad, D. W. and Crain, P. (Eds.) (2019). *The Economics of Integrated Pest Management of Insects*. CABI Publishing.

Onstad, D. W., Bueno, A. D. F. and Favetti, B. M. (2019). Economic thresholds and sampling in integrated pest management. *The Economics of Integrated Pest Management of Insects* 122.

Rossi, R. (2019). The EU fruit and vegetable sector: main features, challenges and prospects. European Parliament Think Tank. Available at: https://www.europarl.europa.eu/thinktank/en/document.html?reference=EPRS_BRI(2019)635563 (accessed 16 April 2021).

Santoiemma, G., Tonina, L., Marini, L., Duso, C. and Mori, N. (2020). Integrated management of Drosophila suzukii in sweet cherry orchards. *Entomologia Generalis* 40(3), 297-305.

Sciarretta, A., Zinni, A. and Trematerra, P. (2011). Development of site-specific IPM against European grapevine moth Lobesia botrana (D. & S.) in vineyards. *Crop Protection* 30(11), 1469-1477.

Sousa, A. R., Parra-López, C., Sayadi-Gmada, S., Barandica, J. M. and Rescia, A. J. (2020). A multifunctional assessment of integrated and ecological farming in olive agroecosystems in southwestern Spain using the analytic hierarchy process. *Ecological Economics* 173, 106658.

Sowell, T. (2014). *Basic Economics*. Hachette, UK.

Standing Committee on Agriculture and Resource Management (1996). *Code of Practice for Management of Queensland Fruit Fly*. Department of Primary Industries, Canberra.

Stern, V. M. R. F., Smith, R. F., van den Bosch, R. and Hagen, K. S. (1959). The integration of chemical and biological control of the spotted alfalfa aphid: the integrated control concept. *Hilgardia* 29(2), 81-101.

Walsh, C. S., Ottesen, A. R., Newell, M. J., Hanson, J. C. and Leone, E. H. (2011). The effect of organic and conventional management programs on apple and Asian pear tree growth, productivity, expenses and revenues in a hot, humid climate. *Acta Horticulturae* 903, 665-672.

Wesseler, J. H. H., Bonanno, A., Drabik, D., Materia, V. C., Malaguti, L., Meijer, M. and Venus, T. J. (2015). *Overview of the Agricultural Inputs Sector in the EU*. Brussels: European Union.

Westigard, P. H. (1979). Integrated pest management of insects and mites of pear. In: D. J. Boethal (Ed.) *Pest Management Programs for Deciduous Tree Fruits and Nuts* (pp. 151-202). Springer, Boston, MA.

Chapter 2

Evaluating the economics of integrated weed management

Pieter de Wolf, Saskia Houben, William Bijker and Koen Klompe, Wageningen Plant Research, The Netherlands

1 Introduction

Integrated weed management (IWM) is a strategy in which multiple tactics are combined for optimal weed control. The tactics include not only preventive options like crop rotation and row distance but also control options like herbicides as well as non-synthetic chemical use and mechanical weeding (Fig. 1). Since it depends on multiple tactics and includes many non-chemical measures, the implementation of IWM is considered to contribute to the reduction of pesticide use.

The EU Horizon2020 IWMPRAISE Project promotes the implementation of IWM by developing, testing and assessing IWM strategies that meet the societal demand for farming practices that minimize environmental risks while maintaining productivity. A core activity in this project is a pre-implementation evaluation of the economics and environmental impact of IWM strategies to see which strategies are most promising in terms of their cost-effectiveness, applicability and environmental impact. The most promising options can then be communicated to farmers and advisors. Understanding the differences in

http://dx.doi.org/10.19103/AS.2021.0098.17

Figure 1 In IWM, multiple tactics to control weeds are integrated (Riemens et al., 2022).

cost-effectiveness, applicability and environmental impact of various strategies is also valuable for policy makers, when working on policies for implementing more sustainable weed management practices in agriculture.

In this chapter, we focus on approaches to the economic evaluation of IWM. Whilst IWM does not necessarily exclude chemical weed management control, it is important to develop effective strategies that depend less on herbicides. The efficacy of chemical weed control strategies in European farming systems is threatened by the development of herbicide resistance and the possible withdrawal of specific active ingredients like glyphosate. This chapter focuses on the economic evaluation of IWM strategies that are less dependent on herbicide use or which seek to replace reliance on herbicides entirely.

Although the development and adoption of novel strategies such as IWM are often first promoted by changes in government policies and regulations, it is important to be aware from the start of their potential economic and practical consequences for practitioners. IWM strategies can have a major economic impact on crop operations, for example, by changing row distances or even the existing crop rotation. It is therefore important to assess the cost-effectiveness of IWM programmes, including taking account of local conditions such as market structure, input costs and on-farm limitations and opportunities such as existing crop rotations. These issues raise questions like, 'What are the costs and benefits of a novel strategy? Does the new strategy affect agronomic and financial risks? How does the IWM strategy and the results of a specific experiment translate to particular conditions on a specific farm?' A farmer needs to know the practical implications and economic consequences of different strategies to decide which can be sustainably implemented, whilst researchers need to be able to decide what strategies have the best potential and therefore which strategies should be developed and promoted.

There are some particular challenges when trying to assess a novel strategy. Data availability is one of the biggest challenges. Data from farm accountancy data networks, for example, are not very useful because they relate to conventional economic performance indicators. The use of data from experiments is also problematic because the economics of an experimental set-up are completely different from the economics of a working farm. For example, experiments use much smaller equipment and much more manual labour for monitoring and sampling than a typical farm. Moreover, in many cases, the design of experiments does not entirely match farming practices. Crop rotation, for example, is a primary weed control measure, but experiments often focus on single crops to reduce the complexity and duration of the experiment. In addition, if an economic evaluation can make a contribution to better (cost-effective) solutions, it has to be done at a very early stage of the research, when no experimental data are available anyway.

Any analysis should get as close to practical conditions as possible while taking into account the specific conditions in different regions. The most promising strategy can then be selected for further examination and development before being implemented in practice. This approach was used in IWMPRAISE, starting with a pre-examination of IWM options to decide which were most promising to address the local challenges that farmers faced regarding weed management.

A starting point is to compare the IWMPRAISE approach with other economic evaluations in the same field of IWM. How influential are the choices of data used, calculation methods and what are the pros and cons of different evaluation approaches? Which requirements should be met when in finding solutions that best fit practice? To find out, the chapter starts with a literature review of ways of evaluating the economics of IWM.

2 Approaches to economic evaluation

A literature review of economic evaluation methods shows that most analyses have been done using cost-benefit analyses, some combined with calculation of net present value (NPV). The economic performance of a new farm strategy is then compared to the performance of alternative strategies, often the existing strategy currently applied in practice. Some evaluations were based on multiple years and most analyses were carried out at field level by calculating input costs and output value. Social values were usually defined at the regional or national level and related to effects on health and biodiversity and the degree of contamination of groundwater and surface water. Most of the studies were done after an experiment using the data from that experiment.

In IWMPRAISE, the economic evaluation has been done at the farming system level at the scale of a normal farm. But what is a normal farm? From an agronomic perspective, it should at least resemble a cropping system that is typical for a specific region. From the economic point of view, it makes sense to define a farming system able to generate a reasonable income over the short term. This typical scale is typically higher than the statistical average of a regional farm population, in which small-size part-time farms are often numerically overrepresented.

When the scale and cropping system of this 'standardised farm' are defined, more detailed data about all inputs and outputs are required. For this farm, current weed management practices are defined in detail to establish a point of reference, while one or more IWM strategies are defined for comparative evaluation. This requires several sources: databases with farm data (e.g. for average yields), interviews with (local) experts like researchers, advisors and farmers, and data relating to inputs and yields from field experiments. Experts are crucial to estimate how a farmer could apply a novel strategy in practice in a particular system.

All information is combined to quantify the reference weed management system and the IWM variant(s). Several economic indicators are calculated, including not only the profitability of the farm but also more detailed indicators such as labour use and costs of machinery, assuming that IWM strategies will affect these indicators. From the perspective of the farmer, it is important not just to know the absolute impact (e.g. IWM will cost 10 h extra labour per ha) but also the relative impact: for a large-scale extensive arable farm, for example, total labour use could be less than 20 h per ha, so 10 h extra is a substantial increase. As pointed out by Pardo et al. (2010), it is also important to have a look at the labour distribution during the year. To explain why and how we developed our evaluation approach, we will describe the economic evaluation undertaken in IWMPRAISE of an IWM programme for winter wheat in Denmark and compare this with two other economic case studies on IWM (Pardo et al., 2010; Vasileiadis et al., 2015).

3 The case study in IWMPRAISE

Crop protection and weed management in particular are not limited to one crop but include crop rotation. An IWM programme affects both direct costs (e.g. herbicide input) and indirect costs (e.g. equipment). For both existing and novel weed management systems, data are collected as much as possible on the farm to stay as close as possible to practice. Information was also provided by local experts in a 2-day visit to the relevant farm to define the context, followed by a discussion to define realistic IWM strategies for the future.

The economic performance of IWM for winter wheat production was evaluated for a standard farm in the Sealand region in Denmark. Grass seed production is the main crop in this area in economic terms. Besides grass seed, farmers mainly grow oilseed rape, cereals and sugar beets. Looking to the future, sugar beet and winter wheat production is expected to become less competitive compared to other European production regions, in particular Eastern Europe. This could lead to a decrease and even complete disappearance of sugar beet production in Denmark, while winter wheat could partly be replaced by spring barley. For this reason, sugar beet was not included in the crop rotation of the representative farm.

In this region, not all land is suitable for grass seed production and the acreage is limited as it is always grown on contract. As a result, many farms have two crop rotations (Tables 1 and 2):

- a rotation with grass seed, oilseed rape and cereals; and
- another rotation with oilseed rape and cereals only.

The farm size required to generate a sufficient farmer income over the short term is 200 ha with rotations of 100 ha each. The grass seed crop red fescue (*Festuca rubra*) was chosen as it is a common grass seed crop in the region.

Table 1 Rotation 1 with a grass seed crop on an average farm in Sealand

Year	Crop	Total ha
1	Oilseed rape	16.7
2	Winter wheat	16.7
3	Winter wheat	16.7
4	Spring barley	16.7
5	Grass seed Red Fescue	16.7
6	Grass seed Red Fescue	16.7

Table 2 Rotation 2 without a grass seed crop on an average farm in Sealand

Year	Crop	Total ha
1	Oilseed rape	25
2	Winter wheat	25
3	Winter wheat	25
4	Spring barley	25

We assumed that winter wheat and spring barley are sold directly at harvest, grass seed yield is stored on the farm and oilseed rape is stored outside the farm to be sold later. The winter wheat straw is bailed and sold, but straw from the other crops is left in the field. The farmer does most of the work by himself, except for the straw bailing, which is outsourced to contractors while, during sowing and harvesting, extra labour is hired. For the labour requirement without contract work, we agreed with local experts to settle for 5-6 h labour per ha.

3.1 Current weed management

Weed management is challenging in both crop rotations, especially in rotation 1. Grass seed production has a very low tolerance for weed infestation, especially for grass weeds, because of the risk of contamination of the cultivated grass seeds with weed seeds. Cleaning grass seeds is costly, and yield is reduced due to the loss of grass seeds from cleaning. The aim of weed management of all crops of rotation 1 is to achieve very low weed numbers during the two growing seasons with grass seed production. Weed management in both rotations is becoming more and more difficult due to widespread herbicide resistance from blackgrass (*Alopecurus myosuroides*) which causes problems in effective weed control in cereals.

Weeds are currently managed mainly by herbicide spraying, including pre-harvest glyphosate applications in oilseed rape and following the second-year harvest of grass seeds (Table 3). Ploughing and seedbed preparation are also part of weed management strategy but are also done for other reasons. In the analysis, we split ploughing and seedbed costs between 'weed management' and 'soil structure', with 50% of the investment and operation costs attributed to weed management. Costs for the use of the tractor and spraying machine are attributed to weed management according to the hours used for weed control. The same goes for spraying a tank mix of a herbicide and another pesticide

Table 3 Time of herbicide spraying in both rotations with the current strategy

	Oilseed rape	Winter wheat	Spring barley	Grass seed year 1	Grass seed year 2
Autumn spray	3	1			
Spring spray		1	1	2	2
Glyphosate at the end of the growing season	Yes				Yes
Ploughing	No	Depending on conditions (weed and soil)	No	No	After harvest

(targeting disease or pest management): the costs are split in proportion between these applications.

3.2 Selecting integrated weed management strategies to compare with current weed management

To compare the economics of IWM, two IWM strategies were designed in accordance with the information collected from interviews with experts and farmers. There was a thorough discussion to decide the two most promising and realistic IWM strategies to apply in local conditions. These were as follows:

- IWM wider rows; and
- IWM delayed sowing.

In the IWM strategy that we called *IWM wider rows*, winter wheat and oilseed rape are sown in wider rows to allow for inter-row cultivation to manage weeds mechanically. This resulted not only in a 60% reduction of herbicide use but also in a requirement for new mechanical weeding equipment for inter-row cultivation. Inter-row cultivation was applied twice a season for both winter wheat and oilseed rape. In winter wheat, weed suppression was also supplemented by the use of a false seedbed (Table 4). Glyphosate is still part of the strategy because, according to the experts, the required quality of the grass seed crop cannot be reached without the chemical control of problematic weeds. Organic grass seed production reflects this since it requires expensive post-harvest cleaning to offset not being able to use herbicides.

The other IWM strategy, *IWM delayed sowing*, follows the principles of delayed sowing by sowing winter wheat 2 weeks later to allow time to prepare a false seedbed prior to the sowing of winter wheat. Modelling predicted a good effect on weed suppression, removing the need for herbicide spraying in the spring (Table 5).

Table 4 Times of herbicide sprays and mechanical weeding actions in the wider row strategy for both rotations

	Oilseed rape	Winter wheat	Spring barley	Grass seed year 1	Grass seed year 2
Autumn spray		1 (band spray)			
Spring spray		1 (band spray)	1	2	2
Mechanical weeding	2	2			
Glyphosate	Yes				Yes

Table 5 Amount of herbicide sprays applied in the delayed sowing strategy for both rotations

	Oilseed rape	Winter wheat	Spring barley	Grass seed year 1	Grass seed year 2
Autumn spray	3	1			
Spring spray			1	2	2
Glyphosate	Yes				Yes

4 Comparing the economics of different integrated weed management strategies

The economic and environmental impact of the two IWM strategies was analysed and compared to the current weed management system by focusing on eight indicators. Since we only discuss the economic evaluation in this chapter, we will not discuss the environmental evaluation. The economic indicators used in IWMPRAISE were as follows:

- *Direct energy use:* This is the fuel use for all field operations. A chemical spray with both herbicides and other pesticides was partly assigned to weed management based on the ratio of herbicides to other pesticides in the spray tank. A spray with only herbicides or mechanical weed control was fully assigned to weed control.
- *Farm profitability (costs per €100 revenue):* Total profit is made comparable with the other indicators by calculating the amount of costs per €100 revenue. This is a relative indicator of the profitability of a farm. All the different costs and revenues are calculated per crop and added up to get a total overview. With this indicator, the financial performance of the system can be assessed.
- *Labour costs for weed management*: An IWM strategy with less input of herbicides may result in the requirement for more labour, for example, for mechanical or hand weeding. To assess this within and across cases, we compared the labour requirements for weeding with the total farm labour requirement (in hours). The division between labour for weeding and non-weeding operations was similar to the calculation of direct energy use.
- *Total costs of weed management in relation to total costs of the farm:* The total costs include yearly costs of machinery, labour costs, herbicide costs and fuel costs.
- *Investment costs for weeding equipment:* A different type of weed management may require other machines compared to current weed management practice. This is especially the case when strategies combine chemical, mechanical and even thermal methods, in which more equipment may be required compared to a fully chemical strategy.

> These investment costs are calculated per system. This is done by using the replacement value and comparing the total replacement value of weeding machines to the total replacement value of all machines present on the farm. The mouldboard plough and the machines for seedbed preparation were assigned to be used 50% for weeding, while the tractors and sprayers were assigned as being used for weeding based on the ratio of hours used for weed control. With this, we created an overview of the extra investment that adoption of a new IWM strategy could require.

These indicators were calculated mainly on basis of local data supported by expert estimations using the business economics analysis (BEA) tool. BEA is used to design, evaluate and compare the business economic effects of different farm strategies. This is an Excel-based tool developed by Wageningen Research for arable and vegetable farming systems. Like most other studies, BEA is based on cost-benefit analysis. What differs from most studies was that we evaluated farm decisions at a crop and enterprise level, including the required investment costs, and that we combined this with expert advice to include important information from practice, as well as using information from another database.

This second database is KWIN-AGV 2018 (Wageningen University & Research, 2018), a reference with quantitative agricultural data concerning Dutch arable and field vegetable farming. It contains current balance calculations for conventional and organic practices for the majority of arable crops and field vegetables grown in The Netherlands. Because a database similar to KWIN was not available in Denmark, we discussed with Danish experts whether the Dutch data were applicable in a Danish context. Fortunately, the cultivation of crops in The Netherlands is very similar to cultivation in Denmark, and where required, we replaced Dutch data with data or estimations from Danish experts.

Perhaps the most important single source of information used in IWMPRAISE was information from experts (researchers, advisors and farmers) who were interviewed about the practical implications of implementing particular practices and about local farming conditions. This process can be characterised as 'expert elicitation', a well-established methodology for synthesising expert opinion to allow for parameterisation and quantifying uncertainty. The technique is a valuable way to support decision-making in uncertain conditions and with limited data available (Schmidt et al., 2017). Expert elicitation makes it possible to gather important information such as what happens if farmers no longer have access to certain crop protection products, for example, what happens if farmers can no longer use glyphosate in grass seed production? The aim of this process is to get as close as possible to practice, both for current practice and for the future IWM strategies. The data for the three strategies were collected primarily on site and transformed to relative values to make the different systems comparable.

The costs and profits were mostly based on local data, with expert estimations and data from KWIN-AGV 2018 where necessary. The overall evaluation consisted of model-based calculations of the economics of weed management strategies in a farm context. The main aim was to get a clear picture of the proportions of costs and profits within the enterprise and the magnitude of impact or change compared to current practice. The results were validated with a sensitivity analysis.

The results from our economic analysis show that both IWM strategies do not perform as well as current practice in economic terms (direct energy input, costs per profit, labour for weeding, total costs of weeding and the investments for weeding machinery) (Fig. 2). However, the performance of the *IWM delayed sowing* strategy is close to standard weed management in economic performance. In IWMPRAISE, we analysed the environmental performance of the three systems as well to show the extra costs in relation to environmental impact. The IWM strategies resulted in a lower environmental impact. Besides the comparison between the three systems, the calculations and indicators show the underlying factors involved, that is, which cost items cause lower performance as well as which elements of novel strategies are positive compared to the standard. In this case study, *IWM wider rows* involves

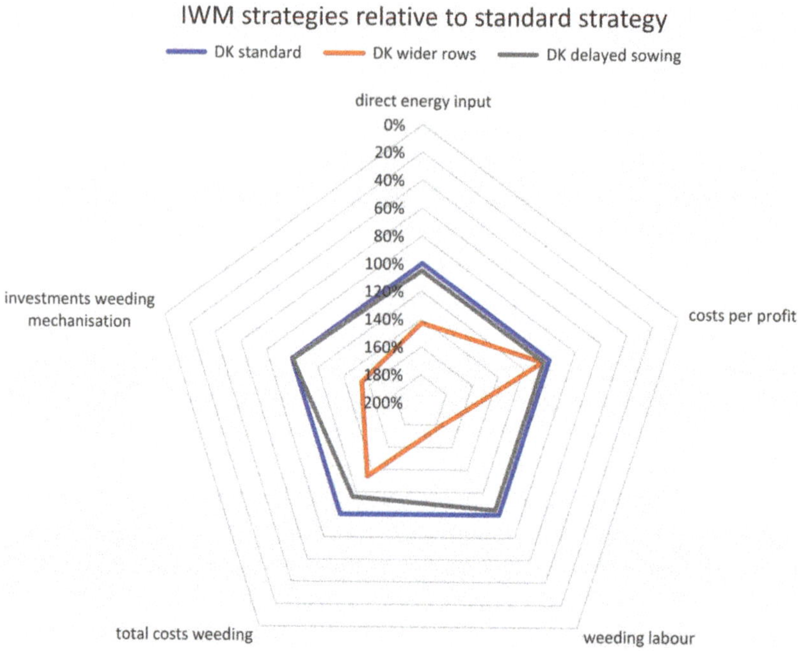

Figure 2 Performance of the two IWM strategies relative to the standard strategy, based on the economic indicators.

mechanical weeding which results in higher fuel consumption and more operations required (e.g. two mechanical weeding operations in oilseed rape with *IWM wider rows* vs. one spray action in standard). Bringing us back to the aim of this evaluation, in terms of economic performance, *IWM delayed sowing* is the most promising strategy for the future when more and more herbicides are banned.

5 Different approaches in assessing the economics of integrated weed management strategies

This section briefly describes two other economic evaluations of weed management which can be compared to the approach in IWMPRAISE (Pardo et al., 2010; Vasileiadis et al., 2015). The approaches of the three studies are compared and summarised in Tables 6 and 7.

5.1 Study by Pardo et al. (2010)

In the study by Pardo et al. (2010), the authors analysed the consequences of several IWM cropping systems both for labour requirements and economic profitability at the farm level. Different IWM cropping systems were analysed using simulations of 10 virtual farms based on a labour requirement model and input data from a 6-year cropping system experiment at a field scale, located at an experimental farm in central-eastern France (Chikowo et al., 2009, after Pardo et al., 2010). One standard weed management system (designed to maximise economic profitability without considering any environmental issues) was compared with four IWM cropping systems. Each of the 10 virtual farms used for model simulations corresponds to one experimental field. A set of decision rules was defined for each cropping system, which was then applied

- to the fields;
- to collect data to test whether the a priori objectives were reached; and
- to perform a multi-criteria assessment of the different cropping systems (Pardo et al., 2010).

The decision rules were based on expert knowledge, simulation with biotechnical models or based on experimental results, either collected locally or reported in the literature (Chikowo et al., 2009; Debaeke et al., 2009; Pardo et al., 2010).

The data collected over the 6 years of experimentation were used to estimate the profitability of the 10 virtual farms, for example, by establishing actual crop yields. Produce selling and input prices were the same as the prices

Table 6 Overview of the three approaches by means of data inputs, analysis methods and tools used

Approach	Input data	Analysis methods and tools
IWMPRAISE	- Direct energy input (fuel use, storage energy use, etc.) - Yields - Weeding labour (relative to total labour) - Total costs of weed management (relative to total costs) - Investments in machines for weed management - Costs per profit	- BEA - KWIN-AGV 2018 - Interviews with experts and farmer - Data from field experiments
Pardo et al., 2010	- Estimations from 6-year field experiment for profitability of 10 virtual farms - Actual crop yields obtained in the experiment - Sale and input prices - Net margin accounting for land rental costs, subsidies, building depreciation, farm overheads, social agrarian insurance - Cumulative time allocated to cultural operations at each time-step compared to the net availability of working time	- Data from a 6-year field experiment --> Estimation of profitability of 10 virtual farms - Ten simulated farms - Trafficable days for each type of operation based on references provided by local agricultural advisory services, climate frequency analysis and model simulations. Validated by technical advisers in the area.
Vasileiadis et al., 2015	- Weed pressure - Crop management cost - Input costs (fertiliser, pesticides, operation costs of labour, mechanical weeding, fuel) - Profit of each commercial or demonstration farm (crop yields, prices)	- Crop yields from all partners involved in experiments - Local prices for contract work (labour, machinery, fuel) - Prices that farmers paid for inputs - Grain maize prices in the years of the experimentation from Eurostat database --> Cost-benefit analysis. - Gross margin: financial yield = (yield × price) – variable costs

on the market at harvesting time in the last year of the experiment. Farmers' income was approximated by calculating the net margin, taking into account public subsidies, rental costs for land, farm overheads, social agrarian insurance and building depreciation.

5.2 Study by Vasileiadis et al. (2015)

As part of the European Pesticide Use-and-risk-REduction in European farming systems with integrated pest management Research Project, Vasileiadis et al.

(2015) evaluated the efficacy of locally selected IWM tools for direct weed control in maize in nine on-farm experiments. They compared these IWM tools with conventional weed management in three different European maize-producing regions. A comparative assessment was undertaken of the economic sustainability of the IWM tools against the conventional weed management approach.

For each farm, the most promising local IWM tools were selected based on

- outcomes of working groups of the EU project (European Network for Durable Exploitation of Crop Protection Strategies) and their results for maize systems in the specific regions (Meissle et al., 2009; Vasileiadis et al., 2011, 2013); and
- consultation with local stakeholders (e.g. farmers, researchers, extension services and agrochemical companies).

The efficacy of individual IWM tools was evaluated by determining weed pressure and through a cost-benefit analysis. Crop management costs (costs of inputs such as fertilizer and pesticides and operational costs such as labour, mechanical weeding equipment and fuel) and crop yields were collected from the experiments. Operational costs were based on local prices for contract work (labour, machinery, fuel) and costs of inputs based on the prices that farmers paid. Grain maize prices over the period of the experiment were derived from the Eurostat database.

6 Comparing different approaches in the economic evaluation of integrated weed management strategies

The approaches in the three case studies have many similarities (Table 7). Each study is based on calculations of costs and benefits to compare the financial profitability of each weed management strategy. Each takes into account the regional setting and uses field data. In the case of IWMPRAISE, however, field data are also used, but costs and profits are calculated within the context of a representative farm, based on its crop rotation and other factors that determine economic feasibility at the farm level.

6.1 The aims of economic evaluation

The aim of IWMPRAISE was to determine whether two specific IWM strategies were a feasible strategy. In addition, we examined what the proportions of different costs were, for example, what were the investments compared to the profits and savings, and how much of the total costs are attributed to weed management? Besides assessing overall feasibility, this evaluation formed a starting point for

Table 7 Overview of the data sources, the results, calculations and methods behind the results

Approach	Source of data			Results	Calculations and methods behind the results
	Experiment/farm	Database or tool	Persons		
IWMPRAISE	- Multiple field experiments - A representative farm	KWIN-AGV 2018BEA	Interviews with local experts and farmers (expert elicitation)	- Economic performance of IWM strategies compared to conventional	Model-based calculations of the costs and benefits of a representative profitable farm
				- Performance on a representative farm (enterprise level)	
				- Investments for weeding and mechanisation	
				- Labour for weeding	
				- Total costs of weeding	
				- Costs per profit	
				- Direct energy input	- Total fuel use at farm level - Fuel use for weed management --> Proportion use for weed management relative to total fuel use
Pardo et al., 2010	- Six-year cropping experiment at an experimental farm at field scale	Simulation model with 10 virtual farms based on data from the 6-year experiment		Economic performance of IWM strategies	Cost-benefit analysis
				- Profitability of 10 virtual farms	Estimated from the data collected in a 6-year experiment
				- Crop yields	Actual yields obtained in a 6-year experiment

				- Selling prices - Input prices	Actual prices at harvesting time in 2006 (last year of the experiment)
				- Net margin	Calculated accounting for public subsidies, land rental costs, building depreciation, farm overheads, social agrarian insurance --> Approximating farmers' incomes
		Climate frequency analysis and model simulations for values used for calculations	- Local agricultural advisory services for values used for calculations - Validation by expert technical advisors from the geographical area	Cumulative time for cultural operations compared to the net availability of working time	- Calculated for each time-step
Vasileiadis et al., 2015	- Nine on-farm (commercial or demonstration) experiments in three regions			- Efficacy of IWM tools for direct weed control in maize compared to the conventional approach - Comparative assessment of their economic sustainability	- Weed infestation in conventional and IWM plots - Cost–benefit analysis
				- Costs of inputs	
				- Operation costs (regional contract work prices including costs for labour, fuel, machinery)	
				- Crop yields and prices	Gross margin = financial yield (yield × price) – variable costs

improvement and adjustment of new strategies to match practical conditions. This approach helps to bridge the gap between research experiments and actual practice with all its regional characteristics and constraints.

A key advantage of the approach used in IWMPRAISE, compared to other studies which used single data from an experiment, was assessing investments at the farm scale. All costs and benefits were made, as much as possible, comparable by calculating relative values, for the following reasons:

- to highlight the balance within the farm, how much is spent on different elements (machinery/labour/investments/permanent costs/weeding, etc.);
- to compare systems across regions/countries/cropping systems/climatic zones; and
- to assess the added value and disadvantages of a novel system compared to a current system.

The time span for assessing economic impact is another important consideration. It is important to calculate NPV to compare similar investment options. Calculating NPV provides insights into the long-term economics of a new strategy, a key concern for farmers or researchers in deciding on which strategy to focus on.

6.2 Data and methods

In IWMPRAISE, data were based on field experiments, a standardised farm, databases and expert elicitation. In the study of Pardo et al. (2010), the analysis was based solely on one experiment. The data were subsequently validated by technical experts and used as input variables in a virtual farm model. Vasileiadis et al. (2015) also used data from just one experiment. The disadvantage of using the data from only one experiment is that it differs from farming practice and gives significant weight to years with either high or low harvest yields. Although the approach used in IWMPRAISE involves more complexity, by combining data from field experiments, databases and expert elicitation, a scenario can be developed that is very close to practice.

Like the IWMPRAISE study, the study of Pardo et al. (2010) stands out from other studies by connecting farm system modelling with experimental cropping system research rather than comparing systems based on cumulative labour requirements for a given crop or rotation alone, which neglects the distribution of labour over seasons and operations.

Pardo et al. (2010), as well as Vasileiadis et al. (2015), expressed the economic impact in absolute values. In IWMPRAISE, we do that by calculating relative values for the different costs and benefits to make a comparison across cases. This brings us back to the question: What do you want to learn from your analysis? If the aim is to show the ratio of costs and benefits to compare strategies, it seems more useful to calculate revenues and reduced costs because that gives a relative difference, making revenues and costs comparable between strategies and across different regions or climates. On the other hand, calculating NPV provides more insight into investment options in considering a new strategy, either as a farmer or a researcher who must decide on which strategies to focus resources.

6.3 Methods for different aims

Vasileiadis et al. (2015) rightfully stated that 'Since all the farming operations and choices, such as crop or hybrid selection, crop husbandry, fertilisation, could affect weed population dynamics, IWM should be fully evaluated not only for its direct weed control but also taking into account the whole cropping system . . . in the long term to capture any shifts in the weed community'. Since crop protection and especially weed management takes place at the crop rotation level, we agree that this should be considered in an economic analysis. Most other studies were not done at the farm level, probably due to the availability of data that are often only from field experiments which do not reflect actual farm practice.

In IWMPRAISE, we used external data to verify whether the data we used were representative of actual farm practice and we used data based on expert elicitation where this brought us closer to practice compared to data from experiments or databases. Other studies do not combine different sources of data in this way. Because we did not depend on experimental data alone, we were able to evaluate the economics of potential IWM strategies prior to doing the research, making it possible to skip research on less-promising strategies and gain insights into possible areas of improvement. The downside is that you always have to stay aware of whether data are 'hard' or 'soft' and which data are based on actual field data and which on expert elicitation? What are the assumptions and what is the possibility that actual implementation will differ from the assumptions made? The same goes for other studies since you have to consider the differences between models and practice, as well as experimental conditions that differ from those in real farms.

To meet the aim of IWMPRAISE - to get insight into the most promising strategies for further development and research - it was logical to evaluate the strategies in the way we did because it brought the evaluation close to actual farming practice and because it evaluated strategies at farm level since IWM involves the entire farm system. For other aims, however, it may make more sense to choose an approach which, for example, is based on experimental data only. Evaluation with multiple sources and expert elicitations is more complex. For example, if the evaluation is conducted for fundamental research, for example, examining the effects of plant physiological mechanisms on productivity, the farm context is not necessarily required. In this case, a strong experimental setup that excludes external factors as much as possible may be more appropriate.

Another consideration, when evaluating the feasibility of a strategy from a farmer's perspective, is the variation in representative farms since each farm is unique, for example, with its own rotation, vision and local conditions. The IWMPRAISE case involving IWM for winter wheat in Denmark involves two crop rotations which represent two systems. The choice of a particular standardised farm setup therefore also affects the result. The context and farm setup should therefore be carefully described so that the differences between strategies can be translated to other situations.

To choose an appropriate approach in an economic evaluation, the following issues need to be considered:

- pre-examination of novel strategies to decide which strategies are most promising to use in practice, for example, to select strategies that will be studied or developed further (like in IWMPRAISE);
- examination of already proven strategies for policy decision-making, for example, to reflect on what is reasonable to demand from the practitioners;
- examination of multiple operation options to show farmers the benefits and disadvantages of each in decision making; and
- regional studies where the aim is to get insight into the effects of a strategy over a time span or sector, for example, public health or economic impact at a system or sector level.

Besides the aim of evaluation, the choice of method depends on the research context and scale of analysis:

- The cost-benefit analysis at the field level (the approach used in the studies of Vasileiadis et al. (2015) and Pardo et al. (2010)) is appropriate for considering costs and benefits and notably to determine the effects of a strategy on the variable costs; and
- Calculations at the farm level, such as IWMPRAISE, are well suited to give insight into the differences between farm systems with differences in investments in new machinery for mechanical weeding. It is therefore necessary to analyse different parameters at the farm level-like crop rotation balance. A cost-benefit analysis at the farm level can also be done but has the disadvantage that it only shows the changed costs and benefits. It does not show whether the economic impact is substantial, for example, an increase of 200 €/ha is significant on a total result of 200 €/ha but relatively minor on a total result of 2000 €/ha.

To conclude, for each approach, there is something to consider. It is important to consider the aim of your evaluation, access to data and the advantages and disadvantages of adding multiple or more complex methods of analysis. For example, expert elicitation gives valuable information about the opportunities, bottlenecks and local conditions in evaluating the economics and feasibility of a system. If the aim is to examine the feasibility of novel measures or strategies from the perspective of a farmer, it is most appropriate to compare a novel IWM strategy with a standardised farm at the farm level, using expert elicitation and data that brings the evaluation close to a real farm situation, including local conditions that can affect a potential solution positively or negatively. By combining multiple data sources and expert elicitation, a strategy can be examined at an early stage and even be adjusted since you can start analysis before experimental data are collected.

7 Where to look for further information

The following guide forms a good starting point to apply economic evaluations of Integrated Pest Management (IPM) specifically:

- Fleischer, G., Jungbluth, F., Waibel, H. and Zadoks, J. C. (1999). *A Field Practitioner's Guide to Economic Evaluation of IPM* (No. 9). Institute for Economics in Horticulture (https://core.ac.uk/reader/29316240).

The following articles provide a good overview of economic evaluation methods, applied in different fields (not specific in agriculture or for IWM):

- Smith, E. G., Upadhyay, B. M., Blackshaw, R. E., Beckie, H. J., Harker, K. N. and Clayton, G. W. (2006). Economic benefits of integrated weed management systems for field crops in the Dark Brown and Black soil zones of western Canada. *Canadian Journal of Plant Science 86*(4), 1273–1279.
- Jones, R., Cacho, O. and Sinden, J. (2014). The importance of seasonal variability and tactical responses to risk on estimating the economic benefits of integrated weed management. *Agricultural Economics 35*(3), 245–256.
- Cameron, N., Wardlaw, T., Venn, T., Carnegie, A. and Lawson, S. (2018). Costs and benefits of a leaf beetle Integrated Pest Management (IPM) program II. Cost-benefit analysis. *Australian Forestry 81*(1), 53–59.
- Visalakshmi, V., Rao, G. V. R. and Rao, P. A. (2005). Integrated pest management strategy against Hubner in chickpea. *Indian Journal of Plant Protection 33*(1), 17–22.
- Gorddard, R. J., Pannell, D. J. and Hertzler, G. (1996). Economic evaluation of strategies for management of herbicide resistance. *Agricultural Systems 51*(3), 281–298.

- Vasileiadis, V. P., Otto, S., Van Dijk, W., Urek, G., Leskovšek, R., Verschwele, A., Furlan, L. and Sattin, M. (2015). On-farm evaluation of integrated weed management tools for maize production in three different agro-environments in Europe: agronomic efficacy, herbicide use reduction, and economic sustainability. *European Journal of Agronomy 63*, 71-78.
- Pardo, G., Riravololona, M. and Munier-Jolain, N. M. (2010). Using a farming system model to evaluate cropping system prototypes: Are labour constraints and economic performances hampering the adoption of Integrated Weed Management? *European Journal of Agronomy 33*(1), 24-32.

The studies in Table 8 can help to decide on a method for economic evaluation.

Table 8 An overview of studies that can be used as an example for economic evaluation

Authors	Title	Level analysis	Methods	Data	Results	Country
Smith et al. (2006)	Economic Benefits of Integrated Weed Management Systems for Field Crops in the Dark Brown and Black soil Zones of Western Canada	Farm level (ha)	• Contribution margin • Statistical analysis to compare IWM with common current practices	• Field experiment • Input: crop price, fertiliser, machinery, seed, marketing	• Means contribution margin	Canada
Jones et al. (2014)	The Importance of Seasonal Variability and Tactical Responses to Risk on Estimating the Economic Benefits of Integrated Weed Management	Farm level (ha)	• Biophysical models • NPV: to compare economic benefit with and without IWM • Stochastic dynamic programming model	• Production inputs • Crop yield and loss • Costs of control • Costs of application • Costs of production • Expert opinions to obtain parameters where no published data was found	• Cumulative probability NPV distributions	Australia

Authors	Title	Level analysis	Methods	Data	Results	Country
Cameron et al. (2018)	Costs and Benefits of a Leaf Beetle Integrated Pest Management (IPM) Program II. Cost-Benefit Analysis[a]	Regional	• Cost-benefit analysis • Cost-benefit ratio	• Costs of control • Yield and loss • Expert opinions to obtain parameters where no published data were found	• NPVs to compare IWM with common current practices for a plantation case study • Comparison of cost--benefit ratios	Australia
Visalakshmi et al. (2005)	Integrated Pest Management Strategy Against *Helicoverpa armigera* Hubner in Chickpea	Farm level (ha)	• Cost-benefit ratio	• Field Experiment • Input: crop price, fertiliser, machinery, seed, marketing	• Yield/ha • Gross income • Costs • Net income	India
Gorddard et al. (1996)	Economic Evaluation of Strategies for Management of Herbicide Resistance	Farm level (ha)	• Cost-benefit analysis • NPV • Economic regression model: MIDAS	• Costs • Yield price • Fixed costs • Variable costs	• NPV/ha • Cost benefits	Australia
Vasileiadis et al. (2015)	On-Farm Evaluation of Integrated Weed Management Tools for Maize Production in Three Different Agro-Environments in Europe: Agronomic Efficacy, Herbicide Use Reduction, and Economic Sustainability	Farm level (ha)	• Cost-benefit analysis	• Experiments farmers' fields • Costs of inputs, e.g., fertilisers and pesticides, and costs of operations, e.g., mechanical weeding, labour, fuel	• Costs/ha • Gross margin/ha	Europe: Italy, Slovenia, Southern Germany

(*Continued*)

Table 8 (*Continued*)

Authors	Title	Level analysis	Methods	Data	Results	Country
Pardo et al. (2010)	Using a Farming System Model to Evaluate Cropping System Prototypes: Are Labour Constraints and Economic Performances Hampering the Adoption of Integrated Weed Management?	Farm level (ha)	• Cost-benefit analysis	• Experiments research	• Input costs comparison • Labour comparison • Income/ha: harvest value • Costs/ha: mechanisation, other inputs, other costs • Net margin	Europe: France

[a]Study is not related to agriculture, but to forestry.

Key research in this area can be found at the following organisation:

- Wageningen Research Field crops (www.wur.nl/en/Research-Results/Research-Institutes/plant-research/Field-crops.htm).

8 References

Chikowo, R., Faloya, V., Petit, S. and Munier-Jolain, N. M. 2009. Integrated weed management systems allow reduced reliance on herbicides and long-term weed control. *Agric. Ecosyst. Environ.* 132(3–4), 237–242.

Debaeke, P., Munier-Jolain, N. M., Bertrand, M., Guichard, L., Nolot, J. M., Faloya, V. and Saulas, P. 2009. Iterative design and evaluation of rule-based cropping systems: methodology and case studies. A review. *Agron. Sustain. Dev.* 7, 235–250.

KWIN-AGV 2018. *Kwantitatieve Informatie Akkerbouw en Vollegrondsgroenteteelt, 2018.* Wageningen University & Research, Business Unit Open Teelten, publicatienummer 776, ISSN 1571-3059.

Meissle, M., Mouron, P., Musa, T., Bigler, F., Pons, X., Vasileiadis, V. P., Otto, S., Antichi, D., Kiss, J., Pálinkás, Z., Dorner, Z., Van Der Weide, R., Groten, J., Czembor, E., Adamczyk, J., Thibord, J. -B., Melander, B., Nielsen, G. C., Poulsen, R. T., Zimmermann, O., Verschwele, A. and Oldenburg, E. 2009. Pests, pesticide use and alternative options in European maize production: current status and future prospects. *J. Appl. Entomol.* 134(5), 357–375.

Pardo, G., Riravololona, M. and Munier-Jolain, N. M. 2010. Using a farming system model to evaluate cropping system prototypes: are labour constraints and economic performances hampering the adoption of integrated weed management? *Eur. J. Agron.* 33(1), 24–32.

Riemens, M., Sønderskov, M., Moonen, A., Storkey, J. and Kudsk, P. 2022. An Integrated Weed Management framework: A pan-European perspective. *Eur. J. Agron.* 133, 126443.

Schmidt, O., Gambhir, A., Staffell, I., Hawkes, A., Nelson, J. and Few, S. 2017. Future cost and performance of water electrolysis: an expert elicitation study. *Int. J. Hydrog. Energy* 42(52), 30470–30492.

Vasileiadis, V. P., Sattin, M., Otto, S., Veres, A., Pálinkás, Z., Ban, R., Pons, X., Kudsk, P., van der Weide, R., Czembor, E., Moonen, A. C. and Kiss, J. 2011. Crop protection in European maize-based cropping systems: current practices and recommendations for innovative integrated pest management. *Agric. Syst. Dev.* 104(7), 533–540.

Vasileiadis, V. P., Moonen, A. C., Sattin, M., Otto, S., Pons, X., Kudsk, P., Veres, A., Dorner, Z., van der Weide, R., Marraccini, E., Pelzer, E., Angevin, F. and Kiss, J. 2013. Sustainability of European maize-based cropping systems economic, environmantal and social assessment of current and proposed innovative IPM-based cropping system. *Eur. J. Agron.* 48, 1–11.

Vasileiadis, V. P., Otto, S., van Dijk, W., Urek, G., Leskovšek, R., Verschwele, A., Furlan, L. and Sattin, M. 2015. On-farm evaluation of integrated weed management tools for maize production in three different agro-environments in Europe: agronomic efficacy, herbicide use reduction, and economic sustainability. *Eur. J. Agron.* 63, 71–78.

Chapter 3

Assessing the economic value of agricultural biodiversity: a critical perspective

Corrado Topi, Stockholm Environment Institute at York, Department of Environment and Geography and Interdisciplinary Global Development Centre, University of York, UK; and Leonie J. Pearson, Stockholm Environment Institute, Thailand

1 Introduction

Over the course of the last few decades, the attempt to give a pure economic value to biodiversity has been a contentious issue and rightly so. Most of the methods that are currently in use at policymaking and decision-making level tend to take a classical economic perspective on the issue of valuing diversity, whether it be biodiversity in general or, more specifically, agricultural biodiversity, that is, agrobiodiversity (terms we will use interchangeably in the rest of the chapter).

The aim of such approaches is to optimise classic economic indicators, such as ROI, which in turn require maximising productivity, that is, taking production per hectare towards the maximum technical potential (MTP), while reducing all costs to the minimum.

This has important consequences for the way in which agrobiodiversity and agriculture are viewed and brings with it a wealth of problems of both an economic and non-economic nature. They range from the ethical impossibility

http://dx.doi.org/10.19103/AS.2020.0071.06

of allocating an economic value to living beings (the main components of biodiversity which includes, of course, humans), to issues of ownership, that is, the fact that biodiversity is a collective good while profit from productivity is an individual good, as well as the error and uncertainty problem, that is, the fact that you cannot price something you do not know, as is the case with many processes that link biodiversity to agricultural productivity. These are but a few of a wealth of similar issues.

The first aim of our chapter is then to highlight some of the most problematic among these many issues.

One answer in applying economic thinking to agriculture is to choose a systemic perspective. A systemic perspective would take into account agriculture within the context of the planetary social, environmental and economic-financial systems, seen as a complex organism made of a series of components, with all their properties, all their relationships and interdependencies (and eventually, as we will see at the end, their rights). Even if a discussion of the properties, structures and theory of complex systems is well beyond the scope of the present chapter, agriculture can be interpreted as a subsystem (i.e. a component), and more specifically, a component of the subsystem that provides food to the human community. By doing so:

- we broaden the system boundary of agriculture to include all of its relationships with all the other subsystems within the planetary boundary,
- we make the constraints imposed by the planetary boundaries on agriculture explicit, and
- we position agrobiodiversity within a much larger map of relationships that may serve other needs (i.e. not agriculture) of the planetary system, accepting the fundamental and intrinsic multidimensionality of the contributions of living beings.

The need to use systemic approaches to make agriculture sustainable is not new to the scholarly community, and we can address readers interested in delving into its historical roots to some seminal work (Ikerd, 1993a,b; Kirschenmann, 1991). However, for a series of reasons, there has been strong resistance to accept that systemic approaches need to be used also in economic analysis of agriculture. One of these reasons is that economics is often confused with accounting which, with its long history dating back to the fifteenth century[1] (Dean et al., 2016; Sangster, 2018), is often considered a well-oiled, completely

1 Living and working in what corresponds to modern day Italy towards the end of the Renaissance, Franciscan friar, polymath and archetypal renaissance man Luca Pacioli is widely regarded in scholarly literature as the 'Father of Accounting' (Fischer, 2000) for his 1494 work *Summa de Arithmetica, Geometrica, Proportioni et Proportionalita* that included a fundamental chapter on theoretical and empirical accounting that de facto gave birth to the discipline (Dean et al., 2016).

settled approach to reality and cost benefits. One other important reason is that within contemporary mainstream neoliberal economic thinking, simplicity, that is, reduction of a system to a small set of easy-to-interpret and formally rigid rules, seems to be a value (inherited from accounting, Sangster, 2018), because it is taken to be formally well founded, and also because it means 'easy replicability', and this is assumed to be possible and true even in the case of complex systems. The rise of systemic science, and of its economic branch, latterly referred to as ecological economics (Costanza, 1991a), is developing instruments for successfully challenging this world view (Costanza, 1991b).

Ecological economics is 'a new transdisciplinary approach that looks at the full range of inter-relationships between ecological and economic systems' where 'by transdisciplinary we mean that ecological economics goes beyond the normal conceptions of scientific disciplines and tries to integrate and synthesize many different disciplinary perspectives' (Costanza, 1991a). As such, ecological economics seems to be the best candidate to deal with systemic phenomena such as biodiversity and agrobiodiversity. Ecological economics also seems to offer the advantages of being a 'post-normal science' (Castro e Silva and Teixeira, 2011), first and foremost its attention to the 'linkages between the social, political, economic, biological, physical, chemical, and geological systems', its explicit claim to ethical non-neutrality, and its 'not being divorced from the policymaking and decision-making processes'. Mainstream orthodox economics comes indeed with a set of implicit underlying ideological assumptions, which, by being implicit, are hard to elicit, challenge and manage. As a consequence, it provides statements with strong ethical implications under the appearance of ethical neutrality, regards linkages mostly as unnatural constraints, and it links policymaking and decision-making only in the form of providing *diktats*[2]. As with all things economic, battle is raging, and the dust is far from settled (Smale et al., 2002), because interests are at stake and world views are being challenged, with repercussions that may go well beyond the realm of agriculture.

The second aim of our chapter is then to critically review developments in methods to assess the economic value of agricultural biodiversity.

The chapter is structured as nine sections, starting with this introduction. In the second section we briefly introduce and discuss the different definitions of agrobiodiversity. In the third section, we discuss ways of evaluating agrobiodiversity. This is followed by Section 4 where we introduce and discuss the ecosystem services framework (ESF), which we extend in the fifth section through the ecosystem interactions framework. In Section 6 we discuss two important limitations, that is, uncertainty and the problem of ecosystem services

2 The assumption of economics, in all of its forms, being an autonomous self-contained 'normal' science at all is a matter of contentious and unresolved epistemological debate.

whose value cannot be derived from the market. We propose an innovative perspective that could radically change the way we evaluate agrobiodiversity in Section 7. Finally, in Section 8 we draw conclusions. The chapter is completed in Section 9 by a series of references for the readers interested in expanding their understanding.

2 The relationship between definitions and economic approaches

In order to clearly identify an economic value for agrobiodiversity, we first need to have a clear definition of agrobiodiversity itself. However, within the literature, there are several, often competing, definitions of agrobiodiversity, which in turn generate very different approaches to its economic evaluation where that is even possible.

Different definitions imply different extrapolation techniques for the different valuation methodologies and for the values themselves and for the valuation of the cost-effectiveness and the benefits of management interventions, when we try to apply them to different situations.

This is especially true for those economic valuation methodologies that focus on indirect estimation and that are, as such, highly context dependent. We use indirect estimation whenever we cannot give a direct value to the real inputs, outputs and inner workings of a process. This requires the use of proxy variables, which in turns increases complexity and uncertainty further. A typical example is using national averages to evaluate the costs of a resource, or of an impact, or the value of a product. Any such transfer of values can thus be unreliable and needs to be treated with caution and adapted on a case-by-case basis.

Several definitions of agrobiodiversity have been given, of which we selected three, representative of the evolution of definitions of the concept, listed in order of time of appearance in the literature:

Definition I (Academic): 'Agricultural biodiversity (or agrobiodiversity) is a component of biodiversity referring to all diversity within and among species found in crop and domesticated livestock systems, including wild relatives, interacting species of pollinators, pests, parasites, and other organisms' (Qualset et al., 1995; Schifino-Wittmann, 2000).

Definition II (FAO definition): 'The variety and variability of animals, plants and microorganisms that are used directly or indirectly for food and agriculture, including crops, livestock, forestry and fisheries. It comprises the diversity of genetic resources (varieties, breeds) and species used for food, fodder, fibre, fuel and pharmaceuticals. It also includes the diversity of non-harvested species that support production (soil microorganisms, predators, pollinators), and those in the wider environment that support agro-ecosystems (agricultural, pastoral, forest and aquatic) as well as the diversity of the agro-ecosystems' (FAO, 1999) (Fig. 1).

Yet, within the literature, and, worryingly, within the more specialised agrobiodiversity economic literature, a different definition is often used:

Definition III (TEEB definition): 'An essential component of the global eco-agri-food system is the genetic diversity of crops and livestock. These genetic resources, including both the diversity of cultivated varieties as well as the wild relatives of crops ("crop wild relatives") and livestock, are a key form of natural capital, and the conservation and use of agrobiodiversity is essential for the development of a more sustainable and resilient global food system' (TEEB, 2018).

In this definition, the system boundary around the agrobiodiversity seems to be essentially restricted to mean genetic diversity of agricultural resources, but misses the contribution of all the wild diversity that provides essential services to the agri-food value chain. Without the latter, the agri-food value chain would have an overall value that is much less, but it is impossible to estimate exactly how large the loss of value would be.

To use the most systemic approach available, we have decided to adopt definition II that is used by UN agencies and that incorporates definition I. We do not use definition III and we suggest this definition is not used in practice. This has important implications for our understanding of how to value biodiversity, as we will discuss in the next section.

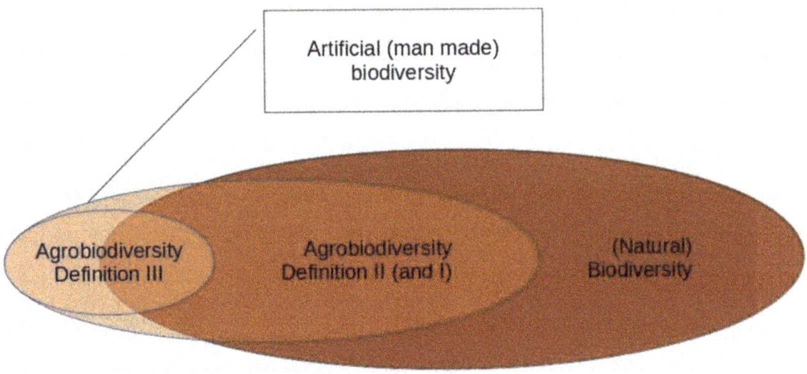

Figure 1 Relationships between the different definitions of agrobiodiversity and biodiversity (natural and artificial).

3 What does valuing agricultural biodiversity mean?

Putting an economic value on biodiversity and on agricultural biodiversity would mean that we can then apply economic abstraction, that can be followed by economic reasoning, on the basis of which decision-makers and policymakers would make (supposedly) rational decisions (Bateman et al., 2013).

Calculating this economic value would require a perfect understanding of the mechanisms by which biodiversity contributes to a human activity. In the case of agricultural biodiversity, decisions on management interventions concerning agricultural land use, including biodiversity, are often either made exclusively on the basis of economic indicators, or at least include economic considerations as a major component of the decision-making process. In the former case, economic abstraction is taken to its extreme, where the object, that is, the agricultural system, is not considered, and its economic value becomes the object of decision-making (as it is the case with some FDI in agricultural systems).

Research that specifically investigates the economic value of agricultural biodiversity is scarce. To validate the claim we conducted a rapid review adapting the methodology of Khangura et al. (2012), where the need was:

- to identify all papers explicitly referring to the economic or financial evaluation of agrobiodiversity, and
- to identify those papers that mentioned economic or financial evaluation of agrobiodiversity in the context of a systemic approach.

We focussed specifically on the Web of Knowledge, one of the most important databases of scientific references available, and, after synonymization and stemming, we formulated and used the query:

*TS=((agr*biodiversity OR (agricultur* NEAR/0 biodiversit*)) AND (economic* OR financ*))*

Our search returned only 251 results on 2020/03/13 at 17:14, while a cumulative search for the last 5 years, and searches by each individual year for the last 5 years did not show a substantial movement away from the yearly average (Fig. 2).

Correspondingly, over the same database and time period, a search for the string:

*TS=((agr*biodiversity OR (agricultur* NEAR/0 biodiversit*)) AND (economic* OR financ*) AND (system* OR integr* OR holist*))*

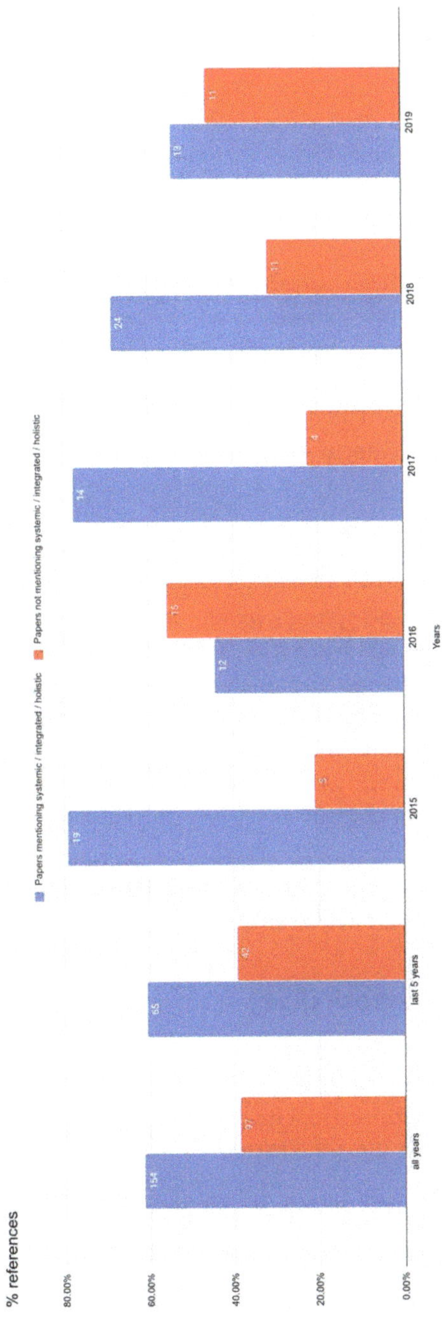

Figure 2 Chart of the % and number of papers on Web of Science mentioning agrobiodiversity economics versus papers that mention it in a systemic study context, 2015–2019, compared to overall average and last 5 years average, from Web of Science. % is given by the height of the column, while the number of papers is given by the label at the top of the individual column.

was run for all years – for the last five years cumulatively and then individually for each one of the last five years. Even this second run did not show again a substantial deviation from the yearly average (Fig. 2).

While this rapid assessment test is far from being conclusive, and must be taken *addito grano salis* (i.e. with a pinch of salt) and only in the context of a rapid review approach, it seems to suggest that using systemic approaches is not yet the norm. More worryingly there is no sign in the literature of a substantial increase or of movement from the year average in the number of papers making references to systemic approaches. While we see a steady number of papers considering biodiversity in agriculture, we see little change from a partial assessment (e.g. consideration of bees as pollinators to specific crops) to a more holistic assessment (e.g. consideration of mixed land use and management, use of native vegetation soil microbes to deliver enhanced agricultural production and ecosystem functioning services such as climate regulation). Without a broader approach, when a 'value' is placed on agricultural biodiversity it is partial, and specific for a given outcome, without considering the whole system.

Systemic farming practices, including those based on agroecological principles, have been delivering sustainable increases in production for many decades with regenerative and semi-closed systems widely seen as the future of agriculture (Pearson, 2007). Yet economics has not kept up with the demand for systemic assessments reflecting these new approaches. In placing a value on agricultural biodiversity, we need to understand the system of assessment – exactly what are we valuing (dollars, outputs or functioning climate), for whom (private landholders or public community), what spatial scale (paddock to landscape) and how (agriculture as a user of biodiversity, or agriculture as part of the biodiversity landscape).

In the next section, we are going to discuss the most important approach that has been devised so far to answer those questions.

4 The ecosystem services framework (ESF)

One of the most significant and important approaches used in numerous studies is the ecosystem services framework (ESF) as a basis to evaluate the economic value of biodiversity. Taking into account the caveats outlined in the previous sections, such an approach has some interesting strengths, even if it is marred by some significant drawbacks.

The idea behind ESF-based approaches is to consider the natural environment as a supplier of services and goods to human economic activities. Within this framework we subsume not only the classic ecosystem services approach, but also the discourse and the debate about natural capital and natural goods, as it is customary.

In theory, such an approach seems to answer directly and specifically the question we are trying to address. If the natural environment can be interpreted as a supplier of goods and services to human economic activities, we can choose to apply 'classic' value chain reasoning and the problem is solved. Unfortunately, this simplistic perspective does not hold true, as we will see in the next sections. As usual, when dealing with truly complex systems, the devil is in the details.

The history of the ESF is several decades long. The framework itself has become mainstream in decision-making and policymaking in the last 20 years, in part as a result of the global effort to produce the Millenium Ecosystems Assessment, also simply known as the MEA (Millennium Ecosystem Assessment, 2005a,b,c,d), where a specific approach to ecosystem services was introduced.

The idea behind this approach was relatively simple. An ecosystem is an extremely complex, intertwined system and the need was felt to capture such complexity within a simplified model. To do so, we can break down the complexity of services offered by the supplier (the ecosystem) into classes and classify all the functions (of which we are aware) into a limited number of homogeneous categories.

On the other hand, we also need to break down the demands of the client (human society) on the supplier (the ecosystem) into similarly homogeneous categories. Finally, we can establish supply and demand relationships between the suppliers' categories on the one hand and the client categories on the other. The results of this conceptual process is captured in Fig. 3 (taken from the MEA (Program), 2005).

Ecosystem services were classified into four major groups: supporting, provisioning, regulating and cultural. Each category or group was then broken down, within certain limits, into a list of services belonging to that group. Of course, a full discussion of the structure of the classification of services in the MEA is outside the scope of the present chapter, and we refer the interested reader to the relevant references about the MEA.

In the MEA approach to ecosystem services, the underlying complexity of functions and relationships relying on biodiversity within the living ecosystem was hidden within the generic category of supporting services rather than being made explicit.

In the early 2000s, policymaking and decision-making bodies started adopting the ESF, including the ESF contained in the MEA, and the language of the ecosystem services (and of natural goods and capital) became progressively more embedded in policy documents.

In parallel to, and following the MEA, several other attempts at classifying the ecosystem services were initiated with a similar world view.

Such classifications and assessments were undertaken or initiated at the global, macro-regional and even national levels, using classifications that

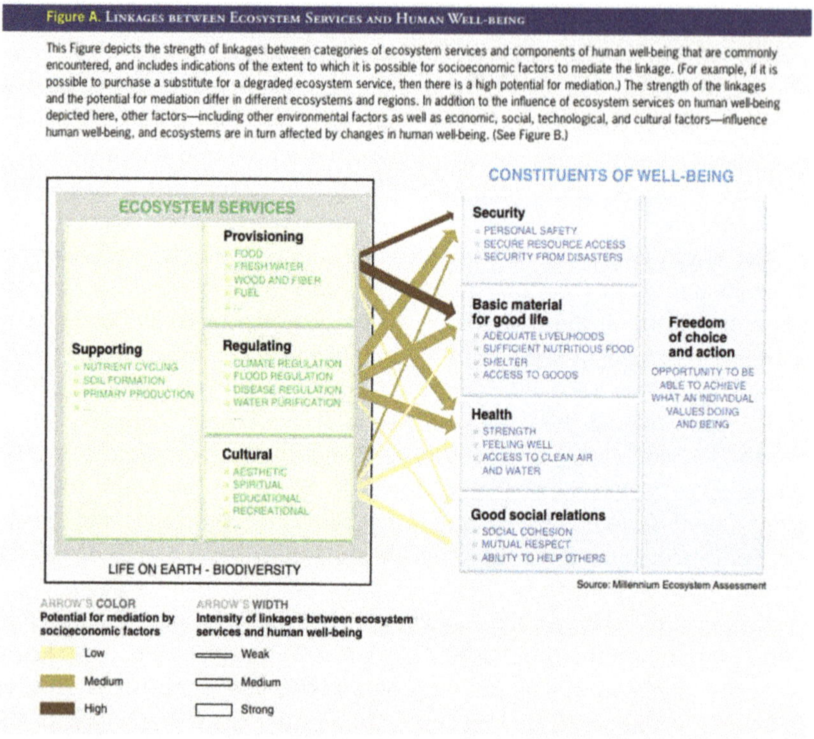

Figure 3 Ecosystem services and their contribution to well-being.

varied, resulting in considerable fragmentation in research, policymaking and decision-making.

While an exhaustive discussion of all the approaches is beyond the scope of this chapter, we think three initiatives are worth mentioning in this context:

- the Economics of Ecosystems and Biodiversity (TEEB) global initiative,
- the Common International Classification of Ecosystem Services (CICES) global initiative, and
- the Mapping and Assessment of Ecosystems and their Services (MAES).

In the following discussion we focus our attention specifically on the attempts at standardisation provided by the TEEB and the CICES initiatives. We leave it to the interested reader to explore the specific macro-regional approach developed by the MAES (https://biodiversity.europa.eu/maes) (Maes et al., 2018). A common feature of all these initiatives is that they consider biodiversity as the basis for the optimal functioning of the ecosystem as a provider of services and goods.

The TEEB global initiative was born out of a proposal of the environment ministers of the G8+5 group at a meeting in March 2007 in Potsdam, Germany.

The TEEB initiative was launched jointly by the German Federal Ministry for the Environment (BMUB) and the European Commission (EC) ('TEEB – The Initiative', n.d.). The aim of the initiative was to analyse the global economic benefits of biological diversity, the costs of the loss of biodiversity and the failure to take protective measures versus the costs of effective conservation. To frame this analysis, TEEB adopted three core principles:

- Recognizing value in all aspects of biodiversity and living ecosystems as a core value of all human societies,
- Demonstrating value in economic terms, considering full costs and benefits of an ecosystem, and
- Capturing value, that is, introducing mechanisms that incorporate the value of ecosystems into decision-making and policymaking.

The work of the TEEB initiative was split into three phases:

- the first phase concluding with the presentation of an interim report at the High-Level Segment of the Ninth Conference of the Parties to the Convention on Biological Diversity (CBD COP-9) in Bonn, Germany, in May 2008 (TEEB, 2008),
- the second phase, concluded with the presentation of four key reports addressed to four different groups of stakeholders in CBD COP-10 in Nagoya, Japan, in October 2010 (Kumar, 2010; TEEB, 2012; T. TEEB, 2009; Wittmer and Gundimeda, 2012), and
- the final current phase, where the findings of TEEB are expected to be embedded into policymaking at all scales following the publication of key studies and of guidelines, including on climate (TEEB, 2009) and the agri-food sector (Ecosystems and Biodiversity, 2018; TEEB, 2018, 2015).

The work of TEEB has been of particular importance for the evaluation of biodiversity and of agrobiodiversity.

The second major global initiative is the initiative by the European Environment Agency (EEA) to sponsor the development of a recognised framework, through the Common International Classification of Ecosystem Services (CICES, https://cices.eu/). As is the case for TEEB, the initiative is currently ongoing. The CICES is now at version 5.1, published in 2018 (Haines-Young and Potschin, 2018, 2013a,b, 2011, 2010).

Compared to other initiatives, CICES introduces two novel elements:

1 it eschews completely the debate around the underlying complexity of ecosystems, by only classifying final ecosystems services, that is, those that provide direct services to humans (Potschin and Haines-Young, 2011), and

2 it includes the abiotic environment in the classification (e.g. climate-related processes), in a provisional state in version 4.3 and in a more systematic way in version 5.1 (Haines-Young and Potschin, 2018)

While the former is concerning, particularly to those that attempt to use systemic approaches for the economic evaluation of (agro)biodiversity, the latter is a welcome addition to the mix.

A very important aspect of CICES is the attempt to harmonise and offer a correspondence between the different frameworks, namely CICES itself, the MEA and the TEEB, and two other classifications not discussed here:

- the Intergovernmental Science-Policy Platform on Biodiversity and Ecosystem Services (IPBES) Classifications, and
- the US-EPA Final Ecosystem Goods and Services Classification System (FEGS-CS).

The CICES makes available simple tools to compare frameworks (https://cices.eu/cices-structure/). The Operationalization of Natural Capital and Ecosystem Services: From Concepts to Real-World Applications (OpenNess) European project provides an online widget to evaluate equivalence between MEA, CICES and TEEB, as well as the UK National Ecosystems Assessment and CICES – Belgium, a nationally adapted version of CICES. While the OpenNess project finished in 2017, the equivalence widgets are available online at http://openness.hugin.com/example/cices and were tested by us on 2020-04-21 at 18:38.

In the next part of this section we review the most common ESF-based methods for valuing agricultural biodiversity based on a review of the literature. While, in the absence of a well-defined attribution protocol, the review makes no claim of being systematic (e.g. as proposed by Sutton et al., 2019 or Khangura et al., 2012), we believe it reflects the major research in the field.

As a starting point Power (2010) has outlined key ecosystem functional relationships within an agroecosystem, assuming that biodiversity is a core input to agroecosystems (Fig. 4). Importantly it recognises both services and disservices that originate from agrobiodiversity.

There are essentially four methods by which the value of agricultural biodiversity can be evaluated in terms of:

1 increasing productivity,
2 increasing resilience,
3 lowering risk, and
4 taking land out of production.

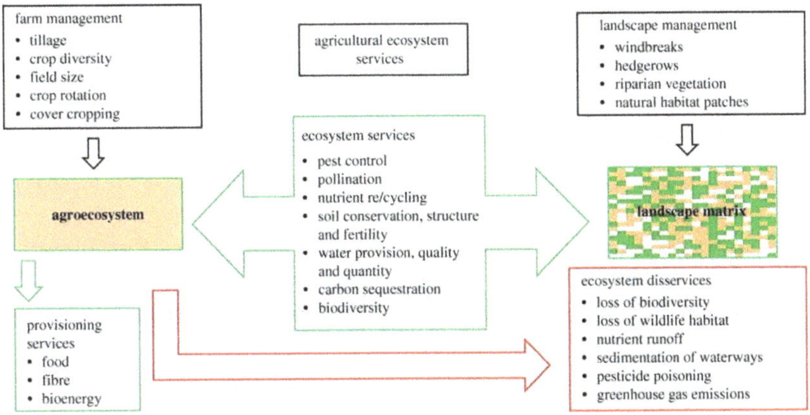

Figure 4 Impacts of agricultural farm management and landscape management on the flow of ecosystem services and disservices to and from agroecosystems. Source: Power, 2010.

These four methods are discussed below.

1 Role of agricultural biodiversity in increasing productivity

A standard economic production function for agriculture shows that, for an increasing production output (Y), there are essential factors of production. This output is only produced, through a functional form (f) of controlled factors like soil, water, seeds and so on (x) and the level of unknown or uncontrollable factors (u).

This can be written as $Y= f(x_1, x_2, x_3....x_n) + u$

Here, we can consider agriculture biodiversity (genetic components) as a factor of production, an (x) – which can have a positive or negative influence on the final outputs (Y). In this way, depending on the type of agricultural biodiversity and its relationship with the output, there are many functional forms to take, increasing linear, decreasing, asymptotic or exponential (see Seppelt et al. (2016) for varying functional forms).

2 Role of agricultural biodiversity in increasing resilience

A specific relationship between agricultural biodiversity and production has been found when biodiversity increases the resilience of the farming system. The value of resilience lies in its ability to maintain the provision of benefits under a given regime. A key assumption is that many farming systems are fragile and vulnerable to a range of intersecting and interacting stresses, for example, flooding, drought, salinization or soil nutrient deficiency. Once certain thresholds are crossed, or when the agroecosystem is 'flipped', a reduction in system resilience would occur, resulting in a decrease in farm productivity.

For example, there is clear evidence that increased soil microbial diversity, quality and quantity can increase farm resilience to drought. Increasing agricultural biodiversity therefore leads to increasing long-term productivity, resulting in increased farm resilience. However, these functional relationships are dependent on the agroecosystem context, level of biodiversity and management techniques (see TEEB 2018 for further references and work on valuation).

3 Role of agricultural diversity in lowering risk

A number of studies have established a relationship between the abundance of biodiversity and the lower risk of crop failure. In particular, there are numerous studies showing that high crop diversity reduces the variance of the yield under varying conditions while radically affecting the mean of the yield. This has been studied both as intraspecific diversity, in particular in cereal production, and in functional farm biodiversity, particularly in modern farms with poor biodiversity (Omer et al., 2007). This relationship seems to apply particularly in more challenging environments, such as more challenging climatic conditions in the developing world, or biodiversity-poor modern agricultural landscapes (see e.g. Di Falco (2012) for a more mathematical approach with a summary of empirical conclusions.

4 Taking land out of production

The relationship between biodiversity conservation and taking land out of production refers to the practice of setting aside a percentage of productive agricultural land in order to foster and promote biodiversity. This may be done in a permanent form, or by rotation, where the areas of the land that are devoted to biodiversity shift periodically, on a percentage of the land at a time, and over periods of time consistent with supporting diversity and with the required rest period needed by the soil. The latter is often one of the most important reasons behind the practice of rotation. Agricultural land that is devoted to biodiversity is taken out of production, which is then typically perceived as a cost of missing revenues. The owner perceives that they incur a pure cost without direct benefit and, for this reason, this is conventionally evaluated economically as a pure cost, that is a lack of productivity from those areas of agricultural landscape that are devoted to biodiversity conservation and cannot be used for direct production.

Additionally, if the owners have to directly manage the land for maximising biodiversity, there may be conversion costs. There are several caveats to this reasoning, in particular, in relation to the fact that productivity is estimated as if the same level of productivity was possible with and without biodiversity.

What are the drawbacks of using the ecosystem services approach in one of its many incarnations to evaluate the economic (and financial) value of agrobiodiversity? In the following discussion we outline four potential weaknesses that should be taken into account when producing any economic evaluation of agrobiodiversity.

The first weakness is that different approaches tend to set system boundaries of varying extensions and coverage, and decisions on what should be included or excluded from a calculation may radically impact the final value. A primary source of concern is the fact that even TEEB, an approach that is in all intentions systemic, puts a rather narrow system boundary around agrobiodiversity as can be seen by the definition of agrobiodiversity given by the initiative (see Definition III in Section 2). Another is that choosing an approach that evaluates only final direct services, while avoiding excessive complexity, does incur the risk of not evaluating correctly the underlying services that are the essential foundation of the final services.

Second, the complex causal and relational connections between the benefits provided by ecosystem services and biodiversity, and in particular agrobiodiversity, are not yet fully understood, particularly for those taxonomic groups that are yet not fully explored, and whose relationship with other taxonomic groups is still to be understood, which makes any overly simplified attempt at economic evaluation very problematic. This is a very important issue because, among those poorly-understood, underexplored taxonomic groups, there are several that are suggested to be of importance for the maintenance of the agri-food supply chain, such as some taxa of insects and of mycorrhizas. Any evaluation should then be used with extreme caution, understood as a provisional qualitative evaluation, and the principle of precaution should always be applied.

Moreover, the role of the abiotic environment in supporting ecosystem services remains unclear, and it is very often wrongly excluded by calculations. We have to remember that even the CICES and ESF did not integrate the abiotic environment until version 5.1, and that even the previous versions, up to version 4.3 that officially preceded the very last release, contained some of the abiotic services in a provisional state.

Fourth, but not least, many of the accounting methodologies based on the ESF appear to capture complexity, but in reality, do not do so, or do so only partially. In particular, they do not take into account five very important aspects of a complex system:

1 transversal interactions between these services,
2 feedback cycles within the ecosystem, within the sphere of the human client and between the two realms,
3 emerging properties, such as pseudo-homeostasis, that are essential to the maintenance of the system as a whole,
4 multi-stakeholder perspectives, conflicts of interests and trade-offs, and
5 tipping points.

While the most recent publications from the TEEB initiative have gone a long way in attempting to include some of these issues, at least conceptually, into consideration of evaluators (e.g. some of the feedback cycles), much is still to be done at a practical level. Discussing this in detail is outside the scope of the chapter. To illustrate our point and give the reader an overview of possible methodological approaches, in Section 5 we look specifically at one of the issues highlighted above: ecosystem interactions.

5 Ecosystem interactions

One of the drawbacks we have highlighted is that the goods and benefits delivered by the ecosystem services in a particular habitat depend heavily on interactions with the adjacent ecosystems and habitats. The current evaluation approaches, when applied tout court (i.e. as they are, without modifying or adapting them) and decontextualized from the actual agricultural landscape and layout of the land, do not take into account the fact that agrobiodiverse agricultural ecosystems behave very differently if interacting with different surrounding ecosystems. An area of farmland devoted to growing cereal crops in a highly agrobiodiverse farm would, for example, behave very differently if the adjacent ecosystems and habitats were rivers and wetlands or forest and shrubland, because the interactions among the ecosystems would be very different. This is particularly relevant for agricultural landscapes, in particular, those characterised by both intensive and extensive farming systems that split the landscape into (relatively) homogenous habitats, such as large adjacent parcels of land for arable for crops, for vegetable growing, tree plantations and so on as well as for those situations where interactions between agricultural habitats and adjacent areas are critical for resilience. For example, different habitats in a landscape can affect each other in the delivery of ecosystem services because there may be trade-offs (negative interactions) and synergies (positive interactions).

This issue is of primary importance and it is generally poorly understood and poorly explored. The statement holds true for all the dimensions: legislation, policymaking, decision-making, conservation, agricultural prices and, last but not at all least, economics.

Recent work for the British government by one of our research groups (Adriana Ford-Thompson et al., 2015) was among the first to explore these interactions using a multidimensional perspective to give practical recommendations for policymaking and decision-making. In the study, we looked at the case study of the complex interactions of agricultural systems, with a specific focus on lowland mixed-use agriculture, urban and fresh water systems.

For the agriculture ecosystems, the study focussed on the production of crops, livestock and fish through aquaculture, while for freshwater ecosystems, the focus was on the specific case of river and wetlands. The case study locations, where the evaluation framework was piloted, were two large multi-habitat regions in the North of England.

We focussed our attention on a subset of all processes and interactions, and more specifically;

1 Category 1: Interactions of agriculture ecosystems with river-wetlands ecosystems:
   1.1. Interaction 1.1: delivery of crops, livestock and fish from aquaculture caused a negative impact upon wild species biodiversity, water quality regulation, water supply, disease and pest regulations and environmental settings in river wetland habitats through pollution.
   1.2. Interaction 1.2: river unmanaged flooding onto agriculture ecosystems caused a negative impact on production of crops, livestock and fish from aquaculture, and on soil quality regulation.
   1.3. Interaction 1.3: river managed flooding for the creation of washlands caused positive impact on wild species diversity, hazard regulation and environmental settings.
   1.4. Interaction 1.4: drainage for agriculture caused positive impact of the production of crops, livestock and fish from aquaculture; negative impact on climate regulation from wetlands by increasing the release of CO_2 and N_2O; positive impact on climate regulation by this reducing emissions of methane; negative impact on trees, vegetation and peat, soil quality and water quality because of peat degradation and of subsidence; positive impact on wild species diversity; positive impact on environmental settings.

2 Category 2: Interactions of urban ecosystems and habitats with river-wetlands ecosystems:
   2.1. Interaction 2.1: pollution from urban areas has negative impacts upon water quality regulation, wild species diversity, disease and pest regulation, water supply and environmental settings in freshwater habitats.

When one looks at the economic evaluation of these interactions, or, more specifically, at the contribution of these interactions to the overall evaluation of the contribution of wild biodiversity, on the one hand, and of agricultural biodiversity, on the other, complexity tends to increase to a point where managing it is not feasible within the constraints of conventional decision-making. We need to adopt an approach that captures part of the complexity, while allowing for some degree of simplification.

Furthermore, we need to include both those processes and interactions whose economic value can be directly evaluated in a conventional economic sense, and those that do not. Among the former are those processes that have or produce or increase benefits and market value, or, on the other hand, destroy, damage or decrease market value and increase costs. Among the latter are those processes that do not have a direct market value.

In the specific case of this research project, we decided to do so by limiting the scope to a series of subprocesses, and then categorising these approaches into broader categories.

1 In the case of Category 1, Interaction 1.2 and, to some extent, Interaction 1.3, the approaches available to economic evaluation can be broadly categorised as 'damage costs avoided'.
2 In the case of Category 1, Interaction 1.4, costs are normally estimated using initial investment, that is, direct drainage costs, and operating costs. They are highly dependent on the stakeholder perspective, for example, for the private landowner the opportunity costs and benefits are a priority, for the community the damage to conservation costs are more important. Opportunity costs and benefits in this context relate to the private cost of agricultural output obtained, or lost if an area is to be destined for conservation for public use, and the operational costs linked to any protective measures adopted. In the latter case, we often need to use non-market value approaches for the benefits, with all the caveats of the case, such as willingness-to-pay approaches, hedonic pricing and travel cost method. We also need to include benefits to the overall community, such as carbon sequestration capacity and recreational spending, and in a limited number of cases, private revenues from activities, such as recreational activities, sports activities, hunting and so on.
3 In the case of both Category 1, Interaction 1.1, and Category 2, Interaction 2.1, the costs and benefits with direct market value can be estimated using replacement costs, substitute values, loss of opportunity costs or loss of earnings, loss of value (e.g. in the case of fisheries), avoided abatement costs, foregone (lost) farm income in relation to establishment of buffer zones and treatment of clean-up costs. In the

case of no market value, these can be estimated using willingness-to-pay and/or willingness-to-accept approaches. These can be based on discrete choice experiments, contingent evaluation and revealed preference methods.

These conclusions, and in particular the methodological framework used, can be easily extrapolated and/or upscaled to cover the cases of other ecosystems and habitats, other landscape scales and other interactions. By doing so, we add a further layer of information to the simpler direct approach offered by ecosystem services applied as such. The outcomes of taking account of interactions in economic assessments may change substantially the way we evaluate the costs and benefits of agricultural biodiversity,

6 Understanding the limitations of ecosystem service valuations

Ecosystem services are not directly traded in any market. Therefore, in undertaking any valuation process for agriculture biodiversity we need to acknowledge that there are some things that are known, but uncertain, others we know we do not know (known unknowns), and others that are unknown unknowns. How we manage the knowns and unknowns is key to determining how robust and transparent agrobiodiversity valuations will be.

As we have seen, there are some things we think we know, for example, how much production a farm typically achieves, what ratio of water, seed, fertilizer and so on is required to make a profit, or what amount of farmland could can be set aside to promote biodiversity (e.g. pollinating species) that would enhance agricultural production. Yet, within these knowns there are unknown factors, which we may label as 'uncertain knowns'. Examples include: How much rain will be delivered this year? What will be the fuel or fertiliser price? Will there be a pest or disease outbreak or locust plague? These are all the variables that we can understand, often do not like, and often we have enough knowledge to make educated guesses or produce (risk) assessments of the likelihood and consequence of their occurrence. Essentially we have two components that are part of this uncertainty. The first is due to the fact we do not know whether these events will happen or not. The second is due to the conceptual or mental models we use to estimate their effects. We can indeed strive to make such guesses, estimates and risk assessments as accurate as possible using a variety of models, yet they will always include a high level of intrinsic uncertainty which goes beyond the fact that any model we use is intrinsically limited. Often, the uncertainty will be larger than the values we can forecast, and the further we try to project into the future, the larger the uncertainty will be. In addition, the uncertainty in the forecast in a situation of high volatility is extremely high and the statistical noise may become higher than the signal. We are living in a situation of extreme volatility, because of the rapidly changing environmental conditions, due to climate change and environmental deterioration, and because of the many side effects such as financial instability, mass migration and pandemics such as COVID-19. Thus, estimating the level of uncertainty, that is very challenging at the best of times and many times may even be impossible, is exceedingly problematic in the current situation.

A second important source of uncertainty derives from the intrinsic inability of our simplified models to match the true complexity of relationships in the ecosystems and from the fact that any model that evaluates the economic value of biodiversity suffers from the well-known 'garbage in garbage out' issues. The fact that we still do not know the dynamics underlying the relationship between ecosystem functioning and biodiversity sufficiently well and the fact we do not have a sufficient amount of data on a sufficient number of cases contribute to the 'garbage in' part of the equation.

There are then the known unknowns. While an exhaustive discussion of all the known unknowns would be out of the scope of the paper, there is one type of known unknowns that is of particular relevance – those values that we know cannot be known in market terms. In particular, to supplement the discussion, we focus on two types of agricultural biodiversity values not derived from the market and not included at all in current ecosystem service valuations. Protected values and non-use values are known to be excluded from almost all valuations.

Protected values are those values based on alternate value systems that do not fit into the utilitarian market approach. Market values are utilitarian, which is one of the primary dangers of applying economic abstraction to living beings: they can be traded-off against each other. We see this in how production functions normally include agricultural biodiversity as one of many equivalent inputs to productivity. In the utilitarian approach, in essence we could decrease biodiversity if we increased another factor, say fertilizer, to achieve the same productive outputs. What happens then if you have fundamentally 'protected' values, those that cannot be traded off (Morris et al., 2020; Pearson et al., 2012)? These protected values are like moral obligations, religious beliefs, love or rules of law, or they may even derive from the application of the precautionary principle or internationally agreed frameworks of inalienable rights, such as the International Bill of Human Rights or the Sustainable Development Goals (SDGs). These values cannot be marginally increased or decreased for trading-off against other outputs. For example, protected values such as clean air and access to clean water for all, or protecting endangered species, may be seen as fundamental moral obligations linked to inalienable rights.

Non-use values are the second type of known value not included in ecosystem valuations generally, but fit within the utilitarian perspective. As part of the total economic value concept, these non-use values encompass the benefits people receive from agricultural biodiversity that are not purely economic. For example, non-use values cover aesthetic value, psychological values of well-being, sense of place and community or values of bequest and benevolence from knowing its existence (see Pearson et al. (2010) for a list of values not derived from the market). In addition, these values may extend beyond individuals who might benefit directly from agricultural biodiversity, for example, people in urban areas, who derive value from the knowledge of agricultural biodiversity in action.

The final group of limitations for ecosystem valuations is 'unknown unknowns'. These include our understanding of both human and ecosystem thresholds. As an example, extreme events such as pandemics may result in significant changes in behaviour and consumer choices, altering the amount and type of value that is placed on agricultural production and ecosystem services. Similarly, work has identified that current global systems may be approaching a threshold, where a return to current functioning is not linear but either asymptotic, exponential or not possible. Such thresholds have been identified both at the planetary scale and at more local agroecosystem scales. The best approach to deal with unknown factors in ecosystem valuation is to clarify the assumptions under which the valuation occurs. Attempting then to undertake a sensitivity analysis to determine how far these unknown changes might affect your valuation may offer some support to decision-making processes. Yet, in the case of systems affected by high volatility, sensitivity analysis requires pushing the boundaries of the envelope of the analysis to extreme cases, where the system may be unable to function: as such they may yield paradoxical results[3].

7 The investor perspective: the natural environment as a legally structured persona

What is next then for biodiversity evaluation, and more specifically, for agrobiodiversity evaluation?

The still incomplete understanding of the complex systemic issues we have outlined in this chapter makes any evaluation of the costs and benefits of biodiversity in general, and of agrobiodiversity in particular, highly approximate, imprecise and affected by uncertainty. It is likely that the true value of the

3 A typical example of a paradoxical result that sensitivity analysis may yield is that of greenhouse gas emissions from human activities decreasing to zero (a 'good' outcome) because everybody is dead and the human species is extinct.

benefits provided by biodiversity and agrobiodiversity is much greater than what we currently estimate.

Within this context, it is important to note that any value that is attributed to biodiversity and agrobiodiversity is negotiated; that is, it is the outcome of a social, political and economic negotiation between different parties such as policymakers, decision-makers, investors, farmers, local communities, non-governmental organisations, practitioners and experts (including economists) representing competing interests.

One option is to give the natural environment a voice and a legally structured persona.

While there is a strong ethical argument for doing so, this would be outside the scope of the present chapter. Thus, in this section we focus on how this could represent a historical shift in the way we value biodiversity and agrobiodiversity towards an approach which would bring us closer to understanding its real value, embedding also many of the elements that are currently missing, including the social considerations.

Giving the natural environment a voice and a legally structured persona would open up the possibility of the natural environment being represented in negotiations through legally appointed representatives. The general concept of this type of representation is well-embedded in legal frameworks across many cultures and economic systems.

Private enterprises and corporations are a perfect case in point. Corporations are abstract entities that do not exist as physical persons, yet they assume an artificial persona, which allows them to be enshrined into law, given the rights and legal obligations under provisions of the law. They are given the opportunity to defend and enforce their legal rights, to execute legally binding economic and financial transactions and to be represented in negotiations by legally appointed representatives.

In this case, this model can be developed further by giving the natural environment fundamental, that is, inalienable, rights, such as the right to exist, or the right to have its fundamental needs met, rights that should be enshrined into a Natural Environment Bill of Rights similar to the framework of the International Bill of Universal Rights[4].

These rights would include the right to be represented legally, to execute legally binding economic and financial transactions through representatives and act as investors with the same rights we give other investors, not least the right to have a voice in how its assets are valued, the right to withdraw

4 The international bill of universal rights (and duties) started with the approval of the Universal Declaration of Human Rights (UDHR) in Paris on 10 December 1948 with the UN General Assembly resolution 217 A, and was completed by the International Covenant on Civil and Political Rights, and the International Covenant on Economic, Social and Cultural Rights which entered into force in 1976.

representatives from unsatisfactory negotiations and the right to the best possible return on investment, to name but a critical few.

This would open up the possibility of the natural environment being fully and independently represented on company boards with the remit to decide on how maximisation of its assets would work, which would include agricultural biodiversity. Not only giving the natural environment a legal persona whose rights are enshrined into a Bill of Rights has two important consequences. First, the natural environment person would be able to represent its stakeholder interests in any negotiation about the value of biodiversity and agrobiodiversity, which, in the case of an economic endeavour would be shareholder interests. Second, it would have the right to be protected from all hostile acts[5].

Such a Bill of Rights would also enable us to bring into the equation the achievement of climate-change targets and SDGs, in particular how fast we achieve them, and what priority they have in the economic agenda.

It must indeed be understood that the natural environment is an investor in all human economic activities, and that some of the assets brought by this investor are essential, such as oxygen for breathing. Thus, these economic activities would not be possible without this investor, who becomes the holder of a 'golden share', which is effectively a controlling share.

Within the context of this change in approach, it is now necessary to develop a social conversation on such Bill of Rights, on representativity and not least on enforceability, that is, on what it means to enforce the rights of this legal persona.

8 Conclusions

It must be understood, first of all, that the value of agricultural biodiversity has so far only been connected to the value of the final goods or services, hence biodiversity has not been given a value per se. Secondly, system boundary matters: the tendency has been, so far, to only value what we can link in causal relationships, that is, what we can clearly understand and map, where we can clearly see cause and effects, and a connection between ecosystem function and ecosystem services and goods is provided. What we understand is only a small part of a very complex whole (in the sense of complex systems), and the true value is likely to be much greater than what we estimate with current approaches. Within this premises, we argue that the following considerations should apply.

First, valuing biodiversity, whether in general or in agriculture, in a pure economic sense is dangerous and should be avoided, as it is the case for all living and life-sustaining systems. Ethical considerations and, as a

5 The authors are indebted to the late Polly Higgins for this fundamental insight.

consequence, social and political decision-making and policymaking should assert the priority of a robust environment for all independently from economic evaluation. This in turn requires a massive shift in our collective mindset and prioritisation system, which has been strongly resisted so far. Yet, under the combined pressure of climate change, global warming, the financial crisis and its geopolitical consequences, inequality and its social consequences at a global level, and of rapidly progressing environmental degradation, it may be the right time to reconfigure our Weltanschauung and mindsets. The idea that an absolute rational economic value per se exists has been proven to be irrational. We argue that some choices and some decisions, in particular those that are linked to the survival of the species as a whole, should be part of a Bill of Rights, beyond the reach of any economic calculation of costs and benefits.

Secondly, a systemic perspective should always be applied, both in the sense that agricultural biodiversity is part of a complex system in a constant flux of change and it should only be investigated through the lens of complex system theory. The idea that we can capture the value of a small fragment of the system abstracted from the rest is unrealistic and very dangerous. It relies on the mistaken assumption that ecosystems are static and function rigidly in a specific well-defined way that will always be the same. There is so far in mainstream economics an intrinsic inability to understand and interpret complexity, and some of the dynamics of the system, for example, non-linear effects, feedback effects, emerging properties and relations of dependency. This makes pure economic valuation, which relies on understanding the system dynamics completely and beyond the boundaries of the economics subsystem, very unreliable. Second and higher-order effects and benefits, indirect effects and benefits, feedback effects and its costs, should always be included, and intangible and immaterial effects should be included in the evaluation. Uncertainty should always be recognised, made explicit in all stages of the calculations, communicated to the stakeholders and its importance should be clearly stressed.

Thirdly, even in those situations where attempting to produce an economic value is mandatory, conventional reasoning in terms of discount rate is dangerous and should not be applied. In other words, the future should be considered to be of the same value as the present. The asteroid thought experiment[6], often used to justify the choice of rather high discount rates, seems

6 In summary, the asteroid thought experiment is often used to argue that we need to maintain a higher than zero discount rate because an asteroid may strike at any time in the future, which is a metaphor for an unknown species extinction level event of planetary magnitude which would affect our ability to live on the planet. It goes without saying that it would affect the ability to perform any economic activity (dead people don't trade). If that is the case, the argument goes, then the present is worth more than the future. In a system hovering around a tipping point, the passing of which may well cause the system to enter a metastable state unfriendly to human life, adopting behaviours that may cause the system to tip over is equivalent to 'inviting the asteroid'. Economic short termism, which essentially consists in considering today (in particular, today's gain) of much higher value than the future, invites and promotes behaviours that may tip the system, already stressed by climate change and environmental degradation, over into the new unfriendly state, hence inviting the metaphorical asteroid.

to be nonsensical in the context of effects on a system that is hovering around tipping points, because it would essentially mean inviting the metaphorical asteroid to hit. As such, ensuring that biodiversity in general, and agricultural biodiversity specifically, is sustained in the future should be a policymaking priority independently of present value in economic terms. This could be seen as an application of the precautionary principle, because at a time of high volatility in a broad sense, where we may be approaching a systemic tipping point at planetary scale, we do not know what could be the future value of key ecosystem functions.

9 References

Bateman, I. J., Harwood, A. R., Mace, G. M., Watson, R. T., Abson, D. J., Andrews, B., Binner, A., Crowe, A., Day, B. H., Dugdale, S., Fezzi, C., Foden, J., Hadley, D., Haines-Young, R., Hulme, M., Kontoleon, A., Lovett, A. A., Munday, P., Pascual, U., Paterson, J., Perino, G., Sen, A., Siriwardena, G., van Soest, D. and Termansen, M. 2013. Bringing ecosystem services into economic decision-making: land use in the United Kingdom. *Science* 341(6141), 45–50. doi:10.1126/science.1234379.

Castro e Silva, M. and Teixeira, A. A. C. 2011. A bibliometric account of the evolution of EE in the last two decades: is ecological economics (becoming) a post-normal science? *Ecological Economics* 70(5), 849–62. doi:10.1016/j.ecolecon.2010.11.016.

Costanza, R. 1991a. Ecological economics: a research agenda. *Structural Change and Economic Dynamics* 2(2), 335–57. doi:10.1016/S0954-349X(05)80007-4.

Costanza, R. 1991b. *Ecological Economics: The Science and Management of Sustainability*. Columbia University Press.

Dean, G., Clarke, F. and Capalbo, F. 2016. Pacioli's double entry – part of an intellectual and social movement. *Accounting History Review* 26(1), 5–24. doi:10.1080/21552851.2015.1129083.

Di Falco, S. 2012. On the value of agricultural biodiversity. *Annual Review of Resource Economics* 4(1), 207–23. doi:10.1146/annurev-resource-110811-114543.

Ecosystems, T.-T.E. of Biodiversity 2018. Measuring What Matters in Agriculture and Food Systems: A Synthesis of the Results and Recommendations of TEEB for Agriculture and Food's Scientific and Economic Foundations Report. UN Environment Geneva.

FAO 1999. Agricultural biodiversity, multifunctional character of agriculture and land conference. FAO, Maastricht, the Netherlands.

Fischer, M. J. 2000. Luca pacioli on business profits. *Journal of Business Ethics* 25(4), 299–312. doi:10.1023/A:1006281415873.

Ford-Thompson, A., Bell, S., Capelli, S., Naylor, B., Raffaelli, D., Ritson, C., Stokowska, C., Timmis, J., Topi, C. and White, P. 2015. Ecosystem interactions ecosystem service interactions among lowland mixed-use agricultural, river-wetland and urban systems (No. NR0151: Ecosystem Interactions). DEFRA, London, UK.

Haines-Young, R. and Potschin, M. 2010. *Proposal for a Common International Classification of Ecosystem Goods and Services (CICES) for Integrated Environmental and Economic Accounting*. European Environment Agency.

Haines-Young, R. and Potschin, M. 2011. Common international classification of ecosystem services (CICES): 2011 update. Report to the European Environmental Agency, Nottingham.

Haines-Young, R. and Potschin, M. 2013a. CICES V4.3 – Revised report prepared following consultation on CICES version 4, August-December 2012. EEA Framework Contract No. EEA IEA/09/003.

Haines-Young, R. and Potschin, M. 2013b. Common International Classification of Ecosystem Services (CICES): Consultation on Version 4, August–December 2012.

Haines-Young, R. and Potschin, M. 2018. Common International Classification of Ecosystem Services (CICES) V5.1 and Guidance on the Application of the Revised Structure, 53.

Ikerd, J. E. 1993a. The need for a system approach to sustainable agriculture. *Agriculture, Ecosystems and Environment* 46(1–4), 147–60. doi:10.1016/0167-8809(93)90020-P.

Ikerd, J. E. 1993b. The need for a systems approach to sustainable agriculture. In:. *Agriculture, Ecosystems and Environment*. Elsevier, pp. 147–60. doi:10.1016/0167-8809(93)90020-P.

Khangura, S., Konnyu, K., Cushman, R., Grimshaw, J. and Moher, D. 2012. Evidence summaries: the evolution of a rapid review approach. *Systematic Reviews* 1, 10. doi:10.1186/2046-4053-1-10.

Kirschenmann, F. 1991. Fundamental fallacies of building agricultural sustainability. *Journal of Soil and Water Conservation* 46, 165–8.

Kumar, P. 2010. *The Economics of Ecosystems and Biodiversity: Ecological and Economic Foundations*. UNEP/Earthprint.

Maes, J., Teller, A., Erhard, M., Grizzetti, B., Barredo, J. I., Paracchini, M. L., Condé, S., Somma, F., Orgiazzi, A. and Jones, A. 2018. *Mapping and Assessment of Ecosystems and Their Services: An Analytical Framework for Ecosystem Condition*. Publications Office of the European Union, Luxembourg, pp. 1–78.

Millennium Ecosystem Assessment 2005a. *Ecosystems and Human Well-Being*. The World Resources Institute, Washington, DC.

Millennium Ecosystem Assessment 2005b. *Ecosystems and Human Well-Being: Biodiversity Synthesis*. The World Resources Institute, Washington, DC.

Millennium Ecosystem Assessment 2005c. *Ecosystems and Human Well-Being: Opportunities and Challenges for Business and Industry*. The World Resources Institute, Washington, DC.

Millennium Ecosystem Assessment 2005d. *Ecosystems and Human Well-Being: Synthesis*. The World Resources Institute, Washington, DC.

Millennium Ecosystem Assessment (Program) (Ed.) 2005. *Ecosystems and Human Well-Being: Synthesis*. Island Press, Washington, DC.

Morris, J., Ensor, J. E., Pfeifer, C., Marchant, R., Mulatu, D. W., Soka, G., Ouédraogo-Koné, S., Wakeyo, M. B. and Topi, C. 2020. Games as boundary objects: charting trade-offs in sustainable livestock transformation. *International Journal of Agricultural Sustainability* 0, 1–24. doi:10.1080/14735903.2020.1738769.

Omer, A., Pascual, U. and Russell, N. P. 2007. Biodiversity conservation and productivity in intensive agricultural systems. *Journal of Agricultural Economics* 58(2), 308–29. doi:10.1111/j.1477-9552.2007.00091.x.

Pearson, C. J. 2007. Regenerative, semiclosed systems: a priority for twenty-first-century. *Agriculture BioScience* 57(5), 409–18. doi:10.1641/B570506.

Pearson, L. J., Kashima, Y. and Pearson, C. J. 2012. Clarifying protected and utilitarian values of critical capital. *Ecological Economics* 73, 206–10. doi:10.1016/j.ecolecon.2011.10.027.

Pearson, L. J., Pearson, L. and Pearson, C. J. 2010. Sustainable urban agriculture: stocktake and opportunities. *International Journal of Agricultural Sustainability* 8(1-2), 7-19. doi:10.3763/ijas.2009.0468.

Potschin, M. B. and Haines-Young, R. H. 2011. Ecosystem services: exploring a geographical perspective. *Progress in Physical Geography: Earth and Environment* 35(5), 575-94. doi:10.1177/0309133311423172.

Power, A. G. 2010. Ecosystem services and agriculture: tradeoffs and synergies. *Philosophical Transactions of the Royal Society of London. Series B, Biological Sciences* 365(1554), 2959-71. doi:10.1098/rstb.2010.0143.

Qualset, C. O., McGuire, P. E. and Warburton, M. L. 1995. *"Agrobiodiversity" Key to Agricultural Productivity*.

Sangster, A. 2018. Pacioli's lens: god, humanism, euclid, and the rhetoric of double entry. *Accounting Review* 93(2), 299-314. doi:10.2308/accr-51850.

Schifino-Wittmann 2000. Agrobiodiversity: characterization, utilization and management. *Plant Pathology* 49(1), 167. doi:10.1046/j.1365-3059.2000.0423f.x.

Seppelt, R., Beckmann, M., Ceauşu, S., Cord, A. F., Gerstner, K., Gurevitch, J., Kambach, S., Klotz, S., Mendenhall, C., Phillips, H. R. P., Powell, K., Verburg, P. H., Verhagen, W., Winter, M. and Newbold, T. 2016. Harmonizing biodiversity conservation and productivity in the context of increasing demands on landscapes. *Bioscience* 66(10), 890-6. doi:10.1093/biosci/biw004.

Smale, M., Már, I. and Jarvis, D. I. (Eds). 2002. The economics of conserving agricultural biodiversity on-farm. Proceedings of the a Workshop. Godollo, Hungary.

Sutton, A., Clowes, M., Preston, L. and Booth, A. 2019. Meeting the review family: exploring review types and associated information retrieval requirements. *Health Information and Libraries Journal* 36(3), 202-22. doi:10.1111/hir.12276.

TEEB. 2008. *The Economics of Ecosystems Biodiversity: an Interim Report*. A Banson Production, Cambridge, UK. Ed. European Communities, Germany.

TEEB. 2009. *TEEB Climate Issues Update*. United Nations Environment Program, Nairobi.

TEEB. 2009. *The Economics of Ecosystems and Biodiversity for National and International Policy Makers–Summary: Responding to the Value of Nature 2009*. Welzel and Hardt, Wesseling, Germany.

TEEB. 2012. *The Economics of Ecosystems and Biodiversity in Business and Enterprise*. Earthscan Publications, UK and New York, NY.

TEEB. 2015. *TEEB for Agriculture & Food: An Interim Report*. The Economics of Ecosystems and Biodiversity and United Nations Environment.

TEEB. 2018. *TEEB for Agriculture & Food: Scientific and Economic Foundations*. UN Environment, Geneva, Switzerland.

TEEB - The Initiative [WWW Document] n.d. *TEEB*. Available at: http://www.teebweb.org/about/the-initiative/ (Accessed on 21 April 2020).

Wittmer, H. and Gundimeda, H. 2012. *The Economics of Ecosystems and Biodiversity in Local and Regional Policy and Management*. Routledge.

Chapter 4

The economics of smallholder farming

David Eagle and Nadira Saleh, Mennonite Economic Development Associates (MEDA), Canada

1 Introduction

The plight of smallholder farmers is a challenging one that the development industry has been focussing on for many years. As shown elsewhere in this book, the challenges of smallholder farming are complex and require a thorough understanding to propose solutions which can improve food security and income levels. A key component to sustainable agriculture is profitability which is driven by the economic context of the farmer and the corresponding economic decisions made by the farmer. If we understand the economics of smallholder farmers, we can understand how they can continue to be profitable and therefore sustainable in their agriculture endeavours. Land, labour, capital and inputs constitute primary economic factors related to smallholder farmer production. Market access relates to economic factors of sales and distribution. These economic factors are present for any agricultural business but the smallholder farmer faces particular constraints that are discussed in this chapter. The decision-making process of smallholder farmers and their sense of agency are important in understanding the economics of the farm model.

http://dx.doi.org/10.19103/AS.2020.0080.02

What are the limitations on revenue generation for smallholder farmers? How can farmers increase yields in the face of competition for water, land, labour while fending off pests and disease and adapting to changing climate? These are significant questions that are explored in this chapter.

The chapter concludes with a case study on commercialized smallholder farmers. It is based upon the experience drawn from an economic development project that built a cassava seed distribution system in Tanzania[1]. These farmers-turned-entrepreneurs are known to the project as Cassava Seed Entrepreneurs or CSEs. While these are commercialized farmers and not subsistence farmers, the experience of these farmers provides relevant, real-world examples of the economic factors presented throughout the chapter.

2 Segmentation

Before economic factors of smallholder farmers can be analysed, there needs to be an understanding of the characteristics of smallholder farmers. The term 'smallholder' farmer implies that the size of land available to the farmer is the primary criteria used in its definition. Indeed, a common definition of smallholder has been generalized to a farmer who had access to less than 2 ha of land. However, in analysis of data collected by the FAO on smallholder farmers across nine countries (Rapsomanikis, 2015) found significant differences in smallholders between countries and those differences relate to a country's corresponding level of development. In addition, it was found that the distribution of farm size was dependent upon factors such as ecological and demographic conditions and economic and technological factors (Rapsomanikis, 2015). The data suggested that establishing a 2 ha threshold for the general definition of a smallholder farmer does not present meaningful analysis across regions and countries. As a result, there is no one universally accepted definition of a smallholder farmer that can be used in economic analysis.

Among literature, various terms are used to refer to smallholder farmer. In her paper, Murphy (2012) referred to 'small-scale farmer' instead of smallholder to recognize that landholding is not the only criterion and to illustrate the relative scale of farmers. In his book on agricultural development, Mellor (2017) described 'small commercial farm' as one of the four types of households with the other three being large-scale/feudal farm, rural non-farm and urban. For the purpose of monitoring sustainable development goal indicators, the FAO proposed a definition of 'small-scale food producers' as those operating land

1 The *Building an Economically Sustainable Seed System in Tanzania for Cassava* is known as '*BEST Cassava*' is a project funded by Bill & Melinda Gates Foundation from 2017-2021. It is being implemented by Mennonite Economic Development Associates (MEDA) in collaboration with partners such as International Institute of Tropical Agriculture (IITA), Tanzania Agricultural Research Institute (TARI) and Tanzania Official Seed Certification Institute (TOSCI).

in the bottom 40% of national land size distribution, the number of livestock in the bottom 40% of national livestock distribution and economic revenue in the bottom 40% of national economic revenues distribution from agricultural activities (FAO, 2018).

What is common among these terms is the fact that smallholder farmers are in many ways marginalized relative to other producers and markets. Criteria used in smallholder farmer definitions often include farm size, livestock holdings, farm revenue, share of crop sold, income reliance on agriculture and share portion of labour contributed by family. A common attribute with smallholder farmers that are particularly rural is that the farm and the household are inter-related.

Segmentation can be based upon these criteria to divide smallholder farmers into useful groupings. Studies by the Consultative Group to Assist the Poor (CGAP) have analysed smallholder farmers to understand distinct profiles of farmer segments to help financial services, government bodies and agricultural development partners to better serve smallholders. In 2013, Peck et al. (2013) presented a segmentation that differentiated smallholders as 'noncommercial' and 'commercial' based on attributes of land size, crop mix (staple vs. cash), level of engagement with markets and access to financial services and technology. Anderson et al. (2019) analysing the smallholders in six African countries used variables such as educational attainment, socioeconomic status, access to emergency funds, mobile phone ownership and occurrence of an unexpected event in the prior 12 months. It concluded that there were three distinct segments: subsisting, commercializing and diversifying. Figure 1 illustrates the difference between the segments. The subsisting segment tends to be rural, relying primarily on agriculture, relatively low income and focussed on household consumption. Commercializing is also rural and relying primarily on agriculture but with higher income and cash crops and slightly larger land size. The diversifying segment may be rural or urban, focussed on agriculture and wage income, relatively higher income and small land size.

Another important dimension of smallholder farmer segmentation that impacts the economics is gender. Women make up approximately 43% of the agricultural labour force of developing countries (FAO, 2011). However, this significant portion of women's labour is often 'invisible' unpaid household labour. One of the gender-based constraints common to agricultural projects is that women neither identify themselves nor are not identified by others for their roles as farmers. Often, traditions, customs and patriarchal institutions push a narrative that farming is dominated by men. Therefore, many women farmers do not have access to extension, training and agricultural support programmes, which are typically gender-blind and focus male heads of household with hopes that this information trickle down. As a result, the economics of women smallholder farmers is often even more challenging than male smallholder

	Subsisting	Commercializing	Diversifying
Location	Rural (70–90%)	Rural (70–90%)	Urban, peri-urban, and rural
Located in urban area[b]	12%	12%	20%
Sources of income	Agriculture and casual labor (often also in agriculture)	Mainly agriculture	Wage or casual labor, their own business, and some agriculture
Income level relative to the other two segments	Relatively lower	Relatively higher though still poor	Relatively higher though still poor
Main income source: Agriculture[c]	75%	85%	37%
Main income source: Wage labor[d]	4%	3%	10%
Agricultural production	For household consumption, with some surplus for sale	Cash crops for sale	For household consumption
Farm business mindset[e]	64%	85%	31%
Sales channel	Retail, cooperative, or middleman. No contract.	Wholesalers or processors. Seldom a formal contract.	Occasional local sales. No contract.
Contract to sell agricultural outputs[f]	4%	12%	3%
Land size relative to the other two segments[g]	Small to medium	Larger	Smallest
Average (ha)	4.2	5.3	3.1
Median (ha)	2.6	3.3	1.8
Financial tools	Some informal	Formal and informal	Mostly formal
Financially included[h]	30%	36%	34%
Use of informal financial services[i]	25%	29%	24%

Figure 1 Sample profiles of smallholder farmer by segments (adapted from: Anderson et al., 2019 p.4).

farmers. Intersectional gender analysis is required to understand demographic segmentation of smallholder farmer populations, as the needs and constraints are different for male and female farmers, depending on their race, ethnicity, location (especially rural, peri-urban, and urban) and age of farmers.

In summary, segmentation analysis highlights some important observations of smallholder farmers. First, there is not one homogeneous smallholder farmer group and therefore there is not a specific definition of smallholder farmer that is universally accepted. Farm size is not the only criteria used to define smallholder farmer. Second, smallholders exist on a spectrum of subsistence/noncommercial agriculture to commercial agriculture. They are complex and likely have multiple sources of cash flows. The benefit of this segmentation is that it makes meaningful analysis based on economic factors such as land, labour, capital, inputs and market access.

3 Land

In economic terms, assets are required to generate income and ultimately wealth. In agricultural economics, land is a critical asset. High-income countries have built economies that leverage formal land ownership structures. In developing countries, which also have a high reliance on smallholder farmers for agriculture production, the policies and formation of land tenure and land markets is a significant and long-standing issue. This has dated back to colonial times and wraps in notions of national sovereignty and independence. But even more recently land tenure has been highlighted as of critical importance to the design of rural development projects with small farmers (FAO, 2002).

Land-tenure rules define the ways in which property rights to land are allocated, transferred, used or managed in a society. Land tenure is often categorized as private, communal or state-owned (FAO, 2002). Private ownership means the assignment of rights to a private party whether it is an individual, married couple, group of people or corporate body. Communal is where a right of commons exists within a community and each member has the right to independently use the land. State-owned land is where property rights are assigned to an authority in the public sector. In many sub-Saharan African countries, the governments maintain ultimate ownership of all the land in the country regardless of the individual ownership by private citizens. In addition, land is often assigned based on customary rights rather than statutory rights. Customary rights or traditional rights are where land is allocated by local customs and traditions which are recognized by village leaders. These informal property customs and laws often lack official state recognition, but may supersede the formal legal rules and processes. For example, a country's laws may declare that men and women have equal rights to hold property and to inherit it, but if cultural norms and practices are in conflict with such laws, the rights of women are likely to be ignored.

Whether it is acquired through formal or informal means, access to land is critical for smallholders to earn incomes and support their families. Smallholder farmers can experience challenges in accessing land. Because of their relatively small-size, they often have little power or influence in the community, thus they typically have access to the worse land conditions/quality or have land located further away with poor access to infrastructure. Land has implications for both women and youth smallholder farmers. Often land is inherited through patrilineal inheritance, descent and inheritance as traced through the male figure. This has profound implications on gendered power relations for smallholder women farmers. Patrilineal inheritance, which is governed by cultural norms, may allow women to access and use land through marriage, but may limit their access and control over the land, what they are able to grow and whether they gain a profit from its sale. Land is also an asset that can be passed down from one generation to the next. The transferring of land to multiple children results in dividing it into smaller land plots. While land tenure and land access is a complex element, the fragmentation of land into smaller plots further challenges the business economics for smallholder farmers.

In summary, the issue of land tenure and land access is a critical one for smallholder farmers to ensure a profitable and productive life. The complexities of ownership models can impact the economics of smallholder farmers because:

1 It can introduce uncertainty in the future access to the land which may impede investment decisions in their agricultural businesses such as inputs, irrigation and modernization.
2 It may reduce smallholder's ability to acquire finance, which is often collateralized with land.

Land ownership with smallholder farmers is most commonly legitimized by so-called informal customary and traditional mechanisms rather than formal, legal documentation (Mellor, 2017). As smallholders look to commercialize and as countries continue to modernize, more formal and efficient land registration procedures will become increasingly important.

4 Labour

Labour is an important factor in any business, but it is especially important for smallholder farmers. Each stage of the agricultural cycle - preparing land, planting, weeding and pruning, harvesting, packing and transporting - requires significant physical labour. As a result, the models of how labour is utilized in a smallholder business - how much labour and how it is paid for - dramatically impacts the economics of smallholder farmers. The amount of labour required for agriculture business is a challenge to quantify in general terms. It varies drastically with conditions of the field, weather and crops planted. However, it is generally thought to be one of the highest cost categories for many smallholder farmers.

Smallholders are generally thought of as resource constrained with poor land quality and low use of inputs. However, the analysis of the FAO data set of smallholder farmers in nine countries by Rapsomanikis (2015) showed that smallholder farmers had higher yields per hectare than large farms. This inverse relationship between scale (plot size) and land productivity has been debated among agricultural economists for years. (For a summary of research see Gourlay et al. (2017).) While there are many potential reasons that might explain the inverse relationship, Rapsomanikis (2015) observed that the high use of family labour is a significant potential contributor. This is due to the over-use of labour, which is often unpaid or significantly underpaid, and results in higher labour intensity for smallholder farmers (Rapsomanikis, 2015). This is especially true for women who often play supportive and laborious roles like weeding, harvesting, managing livestock for smaller animals. In four out of the nine countries examined by Rapsomanikis, women provide more labour on the farm than men. Further, the labour burden of rural women is not always recognized; tasks such as fetching water, firewood, food preparation, childcare and other reproductive work are unpaid, yet are components contributing to a productive farm.

The rationale for using family labour over hired labour is due to economics; there is evidence that smallholders use hired labour but do so at a declining rate as wages go up. In other words, they would use hired labour more if they could afford it. The skew of smallholder farmers to use a higher portion of family labour over hired labour presents a challenge of understanding the impact of labour on the economics of smallholder farming since it is unpaid, especially women's labour, and is often therefore an unrecorded cost as well. When labour costs are viewed as too high, the smallholder farmer may consider a decision to invest in mechanization to lower its labour costs and increase its efficiency and productivity. Smallholders generally have low use of mechanization and technologies such as tractors and mechanical planters and harvesters. Women smallholders notably do not often have access to mechanized technology.

In summary, labour is considered one of the largest cost components in the business. The amount of labour used in agriculture-based businesses varies by crop, land conditions and climatic conditions. The economics of labour also depends on the extent to which farmers employ mechanization and whether they choose to use hired labour or friends and family who may not be paid labour.

5 Capital

Capital has various definitions in economics and finance. In the context of this chapter and as it relates to the case study presented, capital refers to the assets required by a smallholder to operate their farms, to generate their produce and ultimately earn their incomes. Capital is the resource required to produce revenue and it is measured in terms of its monetary value.

While capital is the resource, it is defined by different forms based on its state, source and ownership. Cash is the first form of capital that can be used to buy land, equipment, infrastructure (e.g. fence, warehouse, mill, nursery, water canals etc.) or to buy inputs or labour or even pay to improve access to market. Once assets are acquired such as real estate, equipment or infrastructure that provide competitive advantage to the farmer they are known as capital assets.

Capital is important for the ongoing operations of a business and sufficient capital provides stability to a business to have resiliency during times of unforeseen challenges that would affect production and sales. Capital is even more important for starting a new business or expanding an existing business. Sources of capital include personal savings and income from the farm operations or other businesses. However, smallholder farmers are characterized by low levels of savings and incomes. In addition, household needs present competing demands for income generated by the business. Smallholders are faced with a dilemma, whether to use capital to reinvest in the farm business or for household needs such as food, school fees, medical bills and home improvements. Therefore, smallholder farmers have challenges of having sufficient capital for their business. This is especially true of considered to be part of the 'subsistence' segment who are farming primarily for their own consumption and only selling small surpluses for a bit of income. However, it is also true for smallholders who are in a more 'commercial' segment as they look to grow and increase their incomes.

5.1 Challenge of cash flow

Cash flow is the incoming and outgoing cash based on the operations of the business. The timing of cash flows is a significant component to the operation of an agriculture business since seasonality of crop results in periods of high revenue and cost. The time of receiving agricultural revenue does not coincide with the time of the costs incurred. This creates a mismatch of when cash is available as capital and when it is needed. At times of such mismatch, availability of financing options is a determinant of smallholder farmer economics.

5.2 Challenge of lack of capital solved by financing

Challenges of capital are often filled by borrowing money and other financing options. A smallholder farmer may use savings to buy a water pump or borrow money from an MFI to finance it. There are yet other financing sources like advance payment received from an aggregator or buyer, inputs sold by input supplier in credit terms, or general loan received by re-financing of capital assets (e.g. loan against land, machinery etc.). The challenges of access to finance are well-known in agriculture and especially for smallholders as banks are reluctant to lend to them (Rapsomanikis, 2015). Smallholder farmers are relatively more exposed to 'force majeure' or problems beyond their control than the businesses lenders are used to lending to. Variations in rainfall (too high or too low), disease infestations and market price fluctuations are some force majeure that affect a farmer with good land and good 'track record' as well. Options of insurance exist, but micro-scale insurance products have not experienced commercial success and size yet. Smallholders in the smallest segments performing subsistence farming on very small plots of land have only opportunities to save a little bit at a time to grow very slowly or rely in informal credit from friends and relatives.

There are informal, private money lenders who may be wealthy individuals or distributors/aggregators of crops who can provide some access to credit but often do so with exorbitant fees that take advantage of smallholders (Isaga, 2018). A study in Tanzania and Uganda on responsible agricultural finance found smallholder farmers accessing these channels which they are known colloquially in Tanzania as 'kukata kichwa' which translates as 'cut your head', and in Uganda as 'mbata' which means 'take your blood' as a dramatic indication of their perception of high interest rates and its impact on their business and livelihood (SICGroup, 2017). However, some smallholders still access these channels either because they do not know of other options or the other options are not accessible to them. Descriptions of the different types of credit and analysing the risks of smallholder farming are beyond the scope of this chapter and will be discussed in detail elsewhere in this book.

5.3 Challenge of lack of capital solved by government and NGOs

With capital challenges and access to finance challenges in the private sector, governments and donor-led development programmes have been working on innovative mechanisms to promote investment in new ways to help smallholders. Policy implementations range from establishing a government bank to provide financial services to smallholders, implementing credit guarantee schemes and directed lending schemes. These must work with the private sector to reduce the risk associated with working in the agriculture sector and particularly with the smallholder farmers.

In summary, smallholders are generally constrained by lack of capital, seasonal cash flows and limited access to finance. Both government and private sector programmes that are targeting smallholders need to understand the constraints of the unique situations of their constituents when developing their plans. If there are international development interventions targeting smallholder farmers access to capital and/or finance, they should be focussed on sustainable solutions such as cost sharing with farmers (as opposed to giveaways), working with finance partners to reduce the risk of lending to smallholder farmers or temporary subsidies to stimulate adoption of new practices.

6 Inputs

In general, inputs are resources that are added to the soil and farm operations. The purpose of using inputs is to increase yield and productivity of the farm. Farming the land removes biomass at harvest so to maintain or improve yield nutrients that often need to be added back to the soil. Sometimes these inputs may come from the farm itself. Manure from livestock, for instance, returns nutrients like nitrogen back to the soil. However, many inputs are purchased off the farm and may include agrochemicals such as fertilizer, herbicide and pesticide or may be seeds, irrigation equipment or machinery. The Green Revolution that took place between 1950s and 1960s demonstrated transformational impact on the agricultural sectors of North America, Asia and

South America. However, the use of these inputs and technologies has not been adopted in Africa which is dominated by smallholder farmers.

Reasons why smallholder farmers have refrained from using inputs can be a combination of lack of capital, lack of access and lack of information. The first is related to a lack of capital at the appropriate time that inputs are purchased as mentioned in the previous section. There can be an interest and intention to purchase inputs but at the time to purchase inputs, the farmer may not have the necessary cash. A study by Anderson et al. (2016) of smallholder farmers in Tanzania on intention to purchase inputs found a large discrepancy between intention to save money for inputs and actually saved money for purchase, only about half managed to do so. In their study of the effectiveness of a national input voucher system with smallholder farmers in Tanzania, Hepelwa et al. (2013) found that farmers did not purchase inputs because of low yields which were also highly variable and the perceived high price of fertilizer. Another deterrent is access to the input market since many smallholder farmers live in rural areas and away from market centres where inputs are sold. The lack of information on the improved benefits often stems from customary practices of farmers which results in a lack of awareness of the benefits of certain inputs. For instance, farmers growing vegetatively propagated crops such as cassava and banana are used to share planting materials freely with one another. As a result, for these crops farmers are not used to purchase planting materials or 'seeds'. Whereas other crops such as rice and maize have more established commercial markets and have been used to purchase improved seed.

The solutions to access to inputs have been a priority area for governments, NGO and private sector organizations. Boniphace et al. (2015) suggested policy-level recommendations include prioritizing market infrastructure to improve access and pricing policies to reduce the price to smallholders. While government-policy changes are often long-term affairs, there have been some governments working actively on the issue. For example, the Kenyan government called the National Accelerated Agricultural Inputs Access programme (NAAIAP), which targeted smallholders with one-time input, grants for fertilizer and maize seed with the expectation that they would use a portion of the revenue generated at harvest to purchase the next year's inputs (FAO, 2014). The important attribute is that these grants are not viewed as 'free' but rather with some conditions to repurchase which promotes sustainability. NGO programmes, often in combination with seed companies or input suppliers, have worked on awareness creation programmes. Demonstration plots located close to farmer customers provide opportunities for farmers to see first-hand potential improvements. In other cases, programmes leverage extension services in the country by building the knowledge of agricultural extension officers who then pass the information to farmers.

In summary, inputs provide a significant opportunity for smallholders to improve their yield and productivity. Access to optimum seed and agrochemical input combinations, where and how to access these inputs and proper application has high implication on farmer economics. Access to information through awareness-creation programmes, demonstration plots and extension services can help the farmer understand the benefits which then enable the farmer to make informed decisions on level of investment in inputs.

7 Market access

A 'market' is a concept that is understood intuitively. In general, we conjure an image of stalls selling various wares. In agriculture we often think of stands of fruit, vegetables, spices, meats, cheeses and handicrafts. In this chapter, a market is a broad concept that includes the systems, procedures and actors that combine to form the environment where agriculture products and services are bought and sold. From the farmer's perspective it represents the customer and the corresponding attributes of selling to that customer such as price, transportation, packaging and distribution. It can also mean selling through distribution channels such as wholesalers, supermarkets or off-take markets such as value-added processors. Marketing refers to the activities that a farmer would do to promote the selling of their products and are typically grouped into four categories: what product is being sold, at what price, in what place and distribution and with what type of promotion.

Market access refers to the ability of farmers to access their customers. As an economic concept, if markets are readily accessible then the costs associated with taking the product to market are lower, which is a good thing. In contrast, relatively inaccessible markets tend to have higher cost implications through transportation costs, unfavourable terms with the distributors or customers, for instance. Effective access to markets to sell agricultural products is an important economic factor which can be challenging for smallholder farmers. Even if farmers have increased yield and production, if they do not have markets that can be accessed effectively then they will not realize the economic benefits of their increased productivity. Market access can be different for men and women, especially in contexts where markets are regulated by cultural norms. Some markets are not considered as appropriate places for women and access to transportation may vary. The dynamics of the market whether it is formal or informal can also have dramatic effects on the cost of the smallholder farmer bringing her product to market and on the price that they will ultimately get for their product.

In general smallholder farmers often face disadvantages in accessing the market relative to large-scale farmers. Reasons for this typically include larger geographic distances and distribution costs (being rural and farther from customers and market), smaller production output (which results in less-favourable selling terms) and lack of information on the market prices and distribution options. Solutions that help smallholders overcome barriers to market access are varied. Some solutions may be technology-oriented providing the farmer with better information about markets such as updated prices. For example, interventions focussed on digital market access such as Mercy Corps Agrifin project leveraged a digital platform to provide access for youth in agriculture to markets, finance and input providers (Mercy Corps, 2019). Some solutions may also involve smallholder farmers getting organized together either in formal or informal groups to achieve benefits of scale, share information and distribution costs. However, these organizations should be formed organically with carefully considered mandate and structure to ensure group cohesion, effective trust and management.

Avenues to market can be varied as well and some may have many steps in a distribution chain. For some products a farmer may sell directly to a customer or at the farm gate, the case for many women smallholders. For others, it may get sold to a 'middleman' that acts as an aggregator from multiple farmers, distributor or wholesaler that resells the product either to end customers, processors or supermarkets. Relationships between these actors may be informal, 'loose' relationships based on verbal agreements or no agreements at all. Or, they may be formal relationships with a contract dictating terms such as price, quantity, quality, dates or other factors in the final purchase agreement. Bargaining power and prices for smallholders can vary, as can the selling price received by men and by women for the same product. Social structures may make it more difficult for women smallholder farmers to bargain effectively with male traders for higher prices.

In summary, smallholder farmers may be at a disadvantage in accessing markets due to factors such as low production volumes, geographic distances, high transportation costs and unfavourable selling terms. Leveraging technology platforms and organizing farmers into like-minded groups can be an effective channel for farmers to distribute their products.

8 Decision-making

This chapter began with segmentation of smallholder farmers which gave insight into the heterogeneous nature of smallholder farmers. It also described the complex nature of smallholder farmer businesses as tightly integrated with their household decisions and financial requirements. The farm and the household are often two dimensions of one entity. Next, this chapter described economic factors of land, labour, capital, inputs and market access which all have impact on the farmers' business. The ultimate combined impact of these economic factors depends on the decisions the farmer makes on a continuous basis. This decision-making is critically important in determining the income a farmer realizes; it is how the farmer manages their business.

The decisions a farmer makes depends on which segment they are in, their priorities for the farm output and the future outlook they see for themselves and their crop. Is the priority to secure a source of food for the household as in a non-commercial, subsistence segment? Or is the priority to secure a source of income as in a commercial segment? Or perhaps, the priority is a mixture of the two - some food for household consumption and some surplus to be sold - based on the smallholder's unique situation. These priorities will mean different decisions for many aspects of the farm business. For instance, a farmer will have to decide what crops to produce, how to produce it, how much to produce, what resources to dedicate and when, what inputs to use, when to harvest, where to sell and for what price. Farmers also may have a short-term time horizon looking out in the next 6–12 months. Or they may have a long-term time horizon over multiple years as they plan to move from a subsistence priority to a more commercial priority and expanding their farm size.

Figure 2 illustrates an example of commercial-farming model decision-making environment.

The context of making these decision-making is complex. Agricultural businesses have a high degree of uncertainty which makes it difficult to plan and make decisions. There are factors that are outside of the farmer's control. Weather conditions such as rainfall, flooding and drought cannot be predicted with accuracy and climate change is making these factors even more unpredictable. The attack of diseases and pests and even market prices all ebb and flow from one season to another. A farmer's experience will give them some wisdom in predicting these within some degree of comfort, but it is never with 100% confidence. In the end, a farmer makes some decisions on optimal use of scarce resources several months or years in advance of a harvest based on educated guesses and experience. These decisions are based on risk and directly impact the economic well-being of the smallholder household. In addition, they make these decisions in challenging economic environment with poor information, lack of access to affordable capital and inefficient markets (Rapsomanikis, 2015).

Figure 3 is an illustration of a smallholder farmer in Bangladesh. It illustrates an example of integrated farm and household activities and the complexity of the farm decisions.

It should be noted that the complexity of these decisions and the factors that go into them are similar between subsistence, non-commercial farmers and commercial smallholder farmers. It is tempting to think that subsistence farming might be simpler, but this is often not the case. It is true that some subsistence farmers may not be commercially oriented in the decision-making because of the lack of coordinated thinking between production and marketing or a lack of

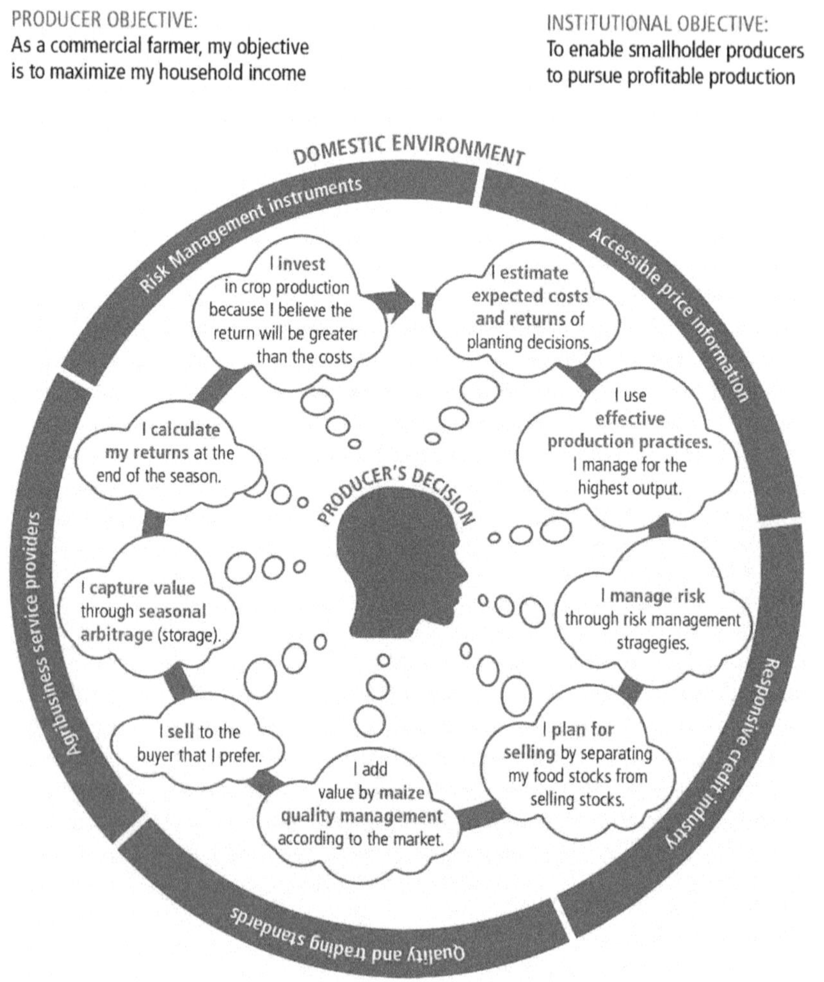

Figure 2 Commercial model decision-making (adapted from: FAO, 2014).

trust in the markets; however, it would be a mistake to generalize all subsistence farmers that way. The difference between subsistence and commercial farmers lies in their differing priorities and context which might result in different decisions. A commercially oriented farmer might decide to sell cash crops with the expectation that the income will be sufficient to both invest capital back to the business and purchase food for their household. Farmers make a series of linked decisions with their unique household circumstances and the local market dynamics in mind. These decisions impact price, quality and distribution expectations. The farmer will then make decisions to organize their farming operations to optimize production and profit (FAO, 2014). Farmers will tend to

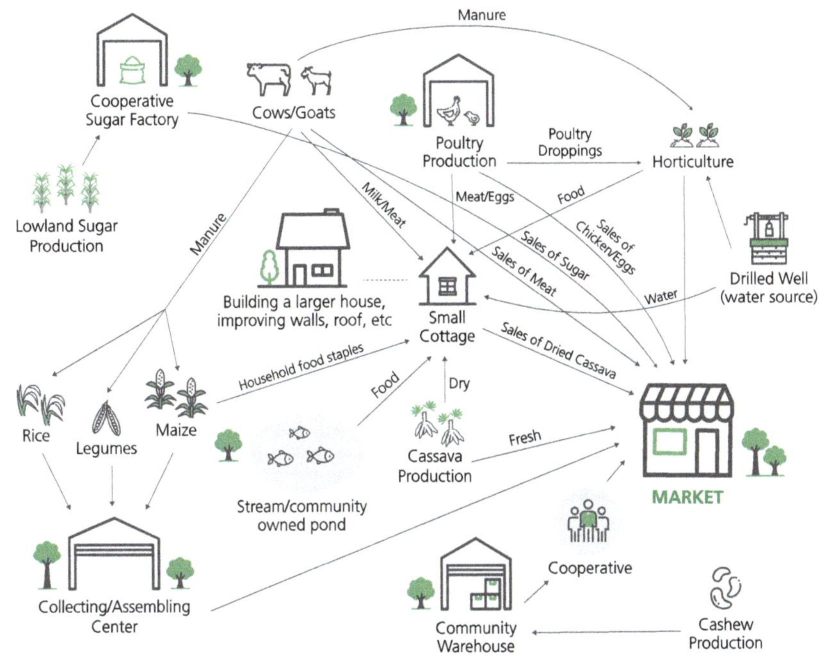

Figure 3 Visual depiction demonstrating decision-making for a smallholder farmer in Tanzania (adapted from: Rapsomanikis, 2015).

make planting decisions based on expected cost and income while managing the perceived risk. Market and price information will be key to these decisions.

In summary, decision-making is critical to farm management. Farmers must make decisions in a complex environment often with relatively high degree of uncertainty. Gathering information helps to manage the risk and makes effective decisions.

9 Case study: BEST Cassava

9.1 Introduction to the case

The case presented here is based on the authors' experience working with farmers in Tanzania growing cassava seed. The Building an Economically Sustainable Seed System in Tanzania for Cassava is known as 'BEST Cassava,' a project funded by Bill & Melinda Gates Foundation from 2017-2021. It is being implemented by Mennonite Economic Development Associates (MEDA) in collaboration with partners such as the International Institute of Tropical Agriculture (IITA), Tanzania Agricultural Research Institute (TARI) and Tanzania Official Seed Certification Institute (TOSCI). This case provides some tangible evidence and illustration of the economic factors impacting smallholders described above. However, it is not presented as an exhaustive case that can provide evidence of all the factors. As has been described above, smallholder farmers are a heterogeneous group making many different decisions based on their own priorities and local context. The economics presented here are representative of the farm business, but analysis is not performed on the household finances overall. Since this project is currently underway, the observations are preliminary. Suggestions for further research and reading are provided at the end of the chapter.

Like many sub-Saharan African countries, Tanzania's economy is dominated by agriculture. Agriculture contributes 28% to the GDP and employs nearly 78% of the workforce (Rapsomanikis, 2015). Of all the farms in Tanzania, it is estimated that 80% are smallholder farms which total 3.7 million farms. Cassava is a vital crop for millions of smallholder farmers in Tanzania but is shifting in status from a survival crop to a staple crop and now to a commercial crop generating an economically sustainable livelihood for farmers. With climate variability, cassava has proven to be among the most resilient crops, lowering the risks for farmers associated with higher-rain-dependency crops. However, pests and diseases cause significant production loss in Tanzania, and the most important of these are the two groups of viruses - the cassava mosaic geminiviruses (CMGs) and cassava brown streak viruses (CBSVs). To date, adoption of clean seed of improved varieties with better disease resistance or tolerance is the most promising way for farmers to combat these problems. Cassava has been prioritized in the new Agricultural Sector Development Plan of Tanzania, not only for its ability to survive in difficult climatic environments, but also as the crop transitions from a staple food to one of increasing commercial value.

Smallholder farmer income and well-being are significantly dependent upon farmer productivity. While some factors such as climate are beyond the farmer's control, access to improved varieties of seed is a large contributing factor to farmer productivity. It is also something that can be improved through an efficient and effective seed production and delivery system. As a result, the potential benefits of improved access to seed through improved productivity and food security are very large. The theory of change for the project to build the cassava seed system in Tanzania is that 'commercially oriented farmers will invest in better seed if it is available and convenient for them to access, if it performs better than what they are now growing, and if the varieties are appropriate for and desired by their local offtake markets'.

Cassava 'seed' production is dependent on a supply of vegetative planting materials (i.e. stem cuttings). The multiplication rate of these materials is very low in comparison to grain crops. In addition, cassava planting materials are bulky and highly perishable as they dry up within a few days after harvest, therefore their multiplication and distribution are more expensive relative to conventional (grain-based) seed services. The yield stability and environmental development of cassava are highly dependent on the quality of planting materials. Cassava cuttings that are continually perpetuated from farmer's own crop year after year tend to accumulate disease and degenerate in yield over time (Shirima et al., 2019). Losses to cassava brown streak disease in Tanzania have recently been estimated at more than US $51 million annually (Ndyetabula et al., 2016).

Since cassava is vegetatively propagated, the stem cuttings are freely distributed between friends and family. This means that farmers traditionally pay very little, if anything, for their planting material. However, it also means that farmers are used to using old cultivars and many already have a disease that gets transmitted with the planting materials. The formal seed system is a deliberately constructed system, which includes a chain of activities leading to clean, disease-free planting materials. The seed system starts with plant breeding and selection, resulting in new varieties and promotes materials leading to formal release and maintenance. Guiding principles in the formal system are to maintain varietal identity and purity and to produce seeds of optimal physical, physiological and sanitary quality.

Cassava seed commercialization is the process of growing or producing cassava seed mainly for profit. The commercialization of cassava seeds is beneficial to ensure a sustainable supply of certified cassava-planting materials and avoids challenges associated with the previous methods of accessing planting materials (free distribution, re-use of own planting material and random exchange of planting material) which increase the spread of diseases, reduce productivity and reach only a limited number of farmers.

The seed system is built upon a network of smallholder farmers that have been recruited and trained to grow cassava cuttings to be used as 'seed'. Profit is a major motivator for smallholder CSEs to produce cassava seeds in a perennial way and at a price that is affordable to farmers for them to purchase. For CSEs at different levels to earn a profit there must be a demand for their product. The seed system has different markets depending on the producer's position in the supply chain. For farmers to purchase planting materials, these seeds need characteristics that suit their needs such as high yielding, preferred taste, disease resistant, early maturity and so on. In the end, there is a mutual benefit between a seed producer and the farmer buying seeds. Nonetheless, despite the potential for profit, there are several considerations for the smallholder, both economic and non-economic that affect her decision-making to become a seed producer.

9.2 Segmentation

The case study presented is based on 349 CSEs across 3 regions, spread over 11 zones in Tanzania. Of these 349, 263 are male (75%) and 68 are female

(25%). Using the CGAP segmentation presented in Section 2 as a guide, the farmers in this case study are considered 'commercializing'. They typically start their cassava seed business with about 3 acres (1.2 ha) of land dedicated to the cassava seed production and may have access to additional land for other crops. Farmers have some business mindset to start and the project activity of recruitment and training helps to build their capacity for record keeping and business-oriented decision-making. The product that they are producing and selling - certified, clean cassava cuttings of improved varieties - is new to the market and requires awareness building through marketing and promotion. Customers are not used to paying much, if anything, for cassava cuttings so there is still some decision-making on the part of cassava seed entrepreneurs to determine appropriate sales prices. Distribution of cassava cuttings is sometimes a challenge due to large geographic distances and the bulky and perishable nature of the product. Significant capital of land and resources are required to start the business as input costs are high for the new planting materials ('seed') and there are additional inspection costs.

9.3 Land

A national survey of smallholders in Tanzania conducted by CGAP in 2016 found that almost half of smallholders (49%) had private control of their land either through lease or deeded ownership. Another 44% had access to the land under customary law (Anderson et al., 2016). For women smallholder farmers, access to land is a barrier as women lack equal rights to inheritance as men under the three systems (statutory, customary and Islamic) of law that govern in inheritance rights in Tanzania. According to SIGI, the local customary law denies widows from inheriting their deceased husband's estates (Tanzania Local Law, 2002). The law also provides for an unequal distribution of property to daughters and sons, allowing for less to be given to the latter (local customary law helping to make clear the disconnection and discord between customary law and statutory law within the context of inheritance rights). While there are statuary laws attempting to limit customary law, such as Probate and Administration of Estate Act Cap. 352 R.E. 2002, section 92 (1), customary law has made it difficult for women and youth to claim and retain their land rights.

None of the cassava seed entrepreneurs involved with the BEST Cassava project own titled and deeded land and instead hold the rights through village allocation systems which are understood to be available to them in perpetuity. Less than 10% of the farmers rent additional land. As a result, the majority of farmers do not show any costs allocated to land in their records.

The reliance on customary rights results in two contrasting realities. On one hand, farmers tend to feel a sense of security and feel comfortable about making decisions about short-term investments. By the same token, few farmers in this business case invest in irrigation or use their land as collateral when seeking to access finance. This is in part due to the cost of such infrastructure, but the lack of permanence of their land and a history of large-scale agricultural investments buying out small farmers may also play a role in their unwillingness to make their own such investments (Sipangule, 2017).

Even without the risk of land seizure, farmers are judicious about how their land is used. Cassava seed entrepreneurs typically have 12 acres of land under cultivation, but the majority limit the size of their multiplication sites to 3-5 acres. There are several reasons contributing to farmers limiting the size of each of their agricultural activities. First, this is a new business and new market and therefore is inherently risky. As a result, farmers still tend to diversify their sources of income.

Another element limiting the growth of CSE farm sizes is the isolation distance that farmers must maintain to grow certified clean seed. Isolation distance, a minimum distance around the cassava field that separating it from other cassava fields, is a requirement by the national seed act for CSEs to pass inspection to certify their stems as disease-free and appropriate for sale. Depending on the type of certification the CSE is applying for, the isolation distance required is between 50 m and 200 m from other cassava fields. The isolation acreage totals approximately 7–20 acres for smaller seed producers. During a 2018 gender assessment, women CSEs and root farmers reported that the isolation distance required was prohibitive to becoming a CSE. Female root farmers reported that the isolation distance required to become a CSE was unattainable for their smaller, packed-together plots. Indeed, some male CSEs reported that to become CSEs, they had converted land from other crops to cassava seed. This resource is often inaccessible to women, limiting their opportunity to become a CSE.

Farmers are able to grow other crops apart from cassava on this land, but in areas where cassava is more common, counter-intuitively, it can be difficult to become a CSE. Smaller farmers with small plots close together face difficulty in achieving the required isolation. Typically, these farmers have agreements with neighbours to respect the isolation distance rather than owning this additional land. Failure to maintain isolation distance can result in possible secondary infections and build-up of the inoculum. The larger the plot of land, the larger the border must be of isolation distance to comply with quality standards for growing quality seeds. This necessitates larger CSEs to be removed from urban centres and away from other cassava producers. For smaller farmers, maintaining the isolation distance is simply not feasible.

Preliminary data analysis from one cycle of harvest and sales shows that even as acreage increases from 5 acres to 20 acres, the opportunity to reach economies of scale is limited – the threshold is much higher to merit introducing mechanized processes like tractors or rototillers. Figure 4 describes changes

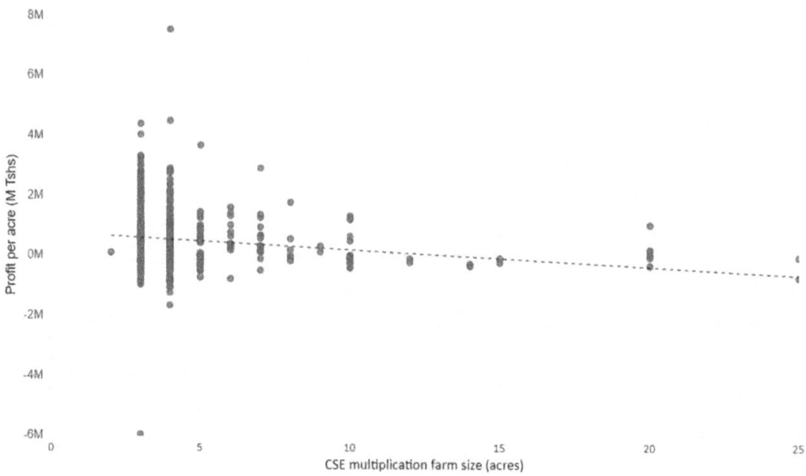

Figure 4 CSE profitability per acre as a function of CSE multiplication acreage.

in profit/acre as acreage increases. There are few CSEs who have larger multiplication sites, making it difficult to draw concrete conclusions about the link between incremental profitability and land size. However, from the data available, it appears that on average, profits remain the same, or decrease slightly as the CSE's plot grows in size.

Without increasing returns, farmers might be reluctant to dedicate too much of their land to one activity.

In summary, land is a critical factor for the CSE businesses. Even though land is allocated without cost based on village allocation/customary rights, the business model that requires isolation distance between cassava fields to be certified is a significant additional burden on the farmers. The model of ownership through village allocation rather than direct-deeded ownership provides some confidence in using the land for their business, but also hinders decisions of investment in technology such as irrigation and hinders the ability to use land as collateral for acquiring additional capital through loans from a formal financial institution.

9.4 Labour

Labour constitutes one of the highest costs for CSEs. In the case of CSEs, it represents 50–70% of direct costs. Most CSEs use a combination of hired and family/in-kind labour for their field activities. However, family labour costs are typically not recorded by farmers as a cost because there is not a cash outlay for it. As a result, the actual amount of labour used in the CSE business may be higher than 50–70%. It should be noted that rural women in agriculture in Tanzania face higher burden than men, regarding responsibilities for unpaid household duties and care work (FAO, 2011). While wages between women and men do not often differ when performing the same activity, in Tanzania women and men tend to have different roles and activities in the field - where men are often responsible for land preparing and planting and women are responsible for cultivation. The activities performed by men tend to have higher wages. The 2018 gender assessment performed during the BEST Cassava project also noted that female CSEs preferred higher amount of female labourers because they are always looking for means to support their families, whereas male CSEs typically hired a higher portion of male labourers over women labourers.

For cassava, as with other crops, labour is most important at the beginning of the planting cycle when farmers must clear, prepare, plant fields and purchase inputs. This imposes high up-front costs, especially for smaller farmers, leading them to resort to personal or family in-kind labour. Figure 5 illustrates with preliminary data the cost of hired labour as a function of acreage of the CSE-multiplication site.

The data demonstrates a nearly flat, slightly decreasing cost for labour costs spent/acre. This indicates that labour costs remain the same per acre for all CSEs regardless of their size; larger CSEs do not get a better deal than those who are smaller. In absolute terms, the data demonstrates that those with more means and larger farms do make use of more labour.

Another factor impacting smallholders' use of labour is its availability. Apart from cassava, other staples like sweet potato and maize also follow the same planting cycle that begin with the onset of the rainy season. With all farmers across a region looking to hire labour at the same time, for those with limited

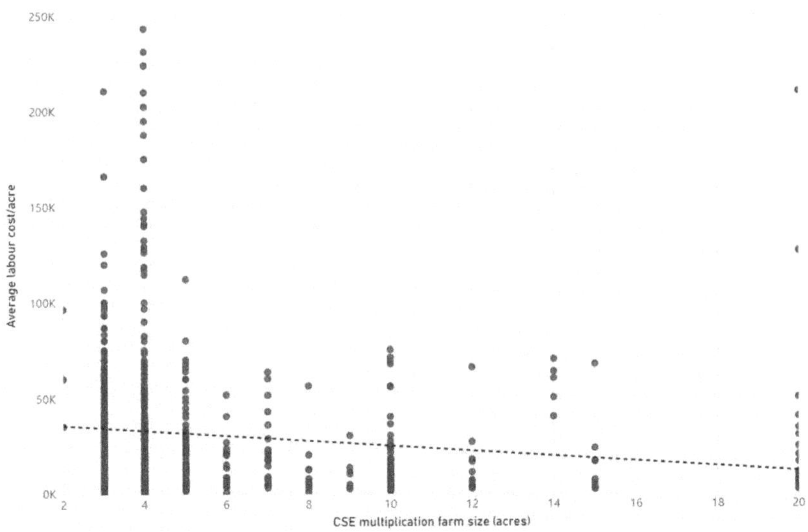

Figure 5 The average labour costs per acre as CSE acreages increase.

means, labour becomes scarce. Farmers also face competition for labour with off-farm income-generating activities; labour is moving increasingly to urban areas as opposed to rural areas. The same labourers who used to be involved in agricultural activities may now work in road construction or brick-making to escape the seasonality of agricultural work. This is especially true for rural youth without any land ownership who are incentivized to take on tasks that pay quickly and reliably. In contrast, agricultural work may have a payment lag, disincentivizing workers.

All this contributes to the fact that CSEs often use in-kind labour. Unfortunately, unlike hired labour, in-kind labour often goes unaccounted by CSEs. Farmers do not track their time or consider it as a cost, so it is difficult for them to understand the true profitability of their operations which may factor in an opportunity cost for their labour. The opportunity cost would be the amount of wages they would receive if they spent that same time working outside their own farm.

In summary, labour was found to be a significant economic factor in the farmer's business. Paid labour represents the majority of the cost (50–70%) of costs to the business. While it is significant, it was not high enough to warrant investment in technologies such as tractors which reduce the cost of labour. However, it was acknowledged that farmers in this case also use in-kind labour from their household which went unaccounted for since the farmers did not consider that a 'cost' of the business.

9.5 Capital

Starting a new cassava seed multiplication business or any new venture for a smallholder can be a risky proposition. Although MEDA's pro forma models indicated reasonably attractive return on investment (ROI) when amortizing start-up costs over three years, many farmers lacked the capital (either cash or credit) to pay the early costs. In the first year, the clearing and cultivation of new land is required, including removal of trees, shrubs, de-stumping all cost-intensive activities. These costs, in addition to purchasing the certified planting material and fertilizer, are required early in the growing cycle and total nearly 30–50% of the total annual costs. Normally, it would take a year before the seed grower would see any revenue generated from selling the stems. This means that a significant portion of the costs need to be laid out up front and the balance throughout the year-long cycle. The harvest, one year later, provides the income needed to offset the costs. During the project gender assessment, women root farmers reported a barrier to becoming a CSE was a lack of steady income and financial resources required as start-up capital. As a result, CSEs are capital-constrained at the outset of the business and the business is risky because they are investing in a new product with unproven market.

Figure 6 highlights a typical cashflow that CSEs experience throughout the growing season. Only after the first year do they record any sales after spending approximately $2020/acre in costs. The time of year this occurs is important: CSEs are able to generate a return (even if it's mostly to recoup their initial costs) during a time of year when they would otherwise be incurring costs for cassava or other crops). After they have harvested their roots, they will have earned nearly double their initial costs ($4200).

To address the issue of lack of capital for farmers the project assisted CSEs in two ways. First, an initial cost-share was provided for 80% of input supplies, including seed and other materials, and 50% of labour and other direct costs paid in cash. This was in recognition of the risk that farmers were taking in starting the new business with scarce resources. Note that the project did not fund outright any activities. For sustainability farmers had to get used to paying the costs and a portion was being recouped from the project. All expenses had to be receipted and no in-kind labour was reimbursed. Altogether, the maximum cost share was approximately US $800 for commercial level CSE.

Year	Month	Activity	Cash Overflow	Cash Inflow
1	August	Slashing and Destumping, Equipment	550	
	September			
	October	Tilling	100	
	November	Ridging	150	
	December	Buying Planting Materials	220	
		Association Fees	170	
	January	Labor	50	
		Transport	30	
		Registration	100	
		Inspection	100	
	February	Fertilizer	180	
	March	Weeding & Thinning	110	
	April			
	May	Weeding	100	
	June			
	July	Weeding	60	
2	August			
	September			
	October	Second Inspection	100	
	November			
	December	Planting Materials Harvest		2,400
	January			
	February			
	March	No Activity		
	April			
	May			
	June			
	July	Tuber Harvest		1,800
3	August			
	September			

Figure 6 Cash flow for CSEs growing both cassava stems (planting material) and roots.

The second mechanism to support capital requirements of CSEs was the creation of a revolving loan fund. This is seen as a longer-term solution that would replace the need for directly subsidized costs. The requirement to create loan funds from scratch within the project stemmed from the fact that most financial institutions are still unwilling to lend to agricultural activities and perceive them to be risky. While Tanzania has established an agricultural bank, it does not operate in several rural areas and the loan criteria for farmers are too stringent for them to access funding. Most farmers in the project have never considered accessing formal loans because they believe they will be ineligible. This may have to do with the fact that historically financial institutions in Tanzania focussed their efforts on corporate customers. One bank was designated for cooperative and rural development (Cooperative and Rural Development Bank - CRDB), but that bank has evolved from that mandate. The name remains the same, but according to the CSEs, the focus has changed to service more affluent and urban customers. With entrance of more banks and increased competition, banks are downscaling their models at regional levels to increase their reach to rural areas. However, this reach is still focussed and preferred at farmer group level or cooperative level lending.

In line with the common 5Cs of lending (collateral, capital, capacity, conditions and character), to access the revolving loan funds, CSEs must fulfil the following criteria:

- Must be a member of seed growers association <Capacity, and Condition>;
- Must have compulsory savings <Collateral>;
- Must contribute shares (equity) to the fund <Collateral and/or Capital>;
- Must have certain activities already completed in their farming process <Capital>;
- Must have two trusted guarantors <Collateral, and Character>; and
- Must have TOSCI Certification <Capacity, and Condition>.

Since the design utilizes savings from all CSEs to provide credit to some, it allows each farmer to make a leap in farm size, capacity and capital base (from savings as a result of higher business yields) during each borrowing cycle. Also, when compared against subsidies that were initially provided, the loan funds serve as a market-based approach, where the CSEs charge interest rates per going rate in the industry (other SACCOs (Saving and Credit Cooperatives) in the vicinity), while being assured that the farmers have access to credit - whereas they may or may not be able to access credit from the SACCOs.

Repayments for these loans have just begun but borrowers seem enthusiastic. Many CSEs have expressed that participation in the fund will help them establish a formal credit history, which can help them access a loan from a commercial bank, especially as they grow. It is expected that these loan funds will be converted into SACCOs by the end of the project.

9.6 Inputs

As seed growers, the farmers in the project purchase clean seed as a regular input to obtain quality certification at the end of the season. Since these are 'seed' businesses they are dependent upon purchasing high-quality cassava cuttings from certified sources to start their businesses. In the proforma model established at the outset of the project, inputs including the seed, fertilizers and herbicides were estimated to contribute 25% of the total costs in year 1. These were approximately evenly split between the seed and chemical inputs. The business model allowed for the seed to be harvested over three annual cycles before being uprooted and planted with new seed. The spreading of the cost of the seed inputs over 3 years is a significant factor in the ongoing profitability of the business.

However, the experience with the CSEs shows that the use of inputs like fertilizer and herbicides is limited. Out of nearly 350 seed growers, only 30 have ever used fertilizers and pesticides. The project regularly recommends the use of these to increase yields and improve performance, but adoption is limited. This leads to an interesting observation of the decision-making process of these farmers. Why did they decide not to use fertilizers and pesticides when it was recommended as best practice? One reason was that fertilizer use for subsistence crops like cassava is often unspecified; there is a blanket recommendation for all crops, but many farmers are not sure of the specific needs for cassava. One of the findings from the project's gender assessment was that agronomy trainings offered by district governments are not always accessible to women, creating a gap between the level of skills and knowledge women may have in agronomic practices. Moreover, there was a perception among farmers that cassava does not require significant inputs. As a hardy root crop, most farmers assume that it will flourish as long as rains are consistent. Lastly, as the project is relatively new and continues to recruit more first-time CSEs, there is an impression that the land being used to grow cassava is 'virgin' and has not yet been depleted of nutrients. Thus, the project may see increased fertilizer use with time.

9.7 Market access

The building of a cassava seed system in Tanzania required recruiting CSEs at various levels in a hub-and-spoke model. 'Basic' level CSEs sold primarily to 'Commercial' CSEs who sold to 'QDPM' CSEs who then sold to farmers growing cassava roots (see Fig. 7). This means that each CSE had to intentionally think of creating a new market. On the positive side, because of the hub-and-spoke

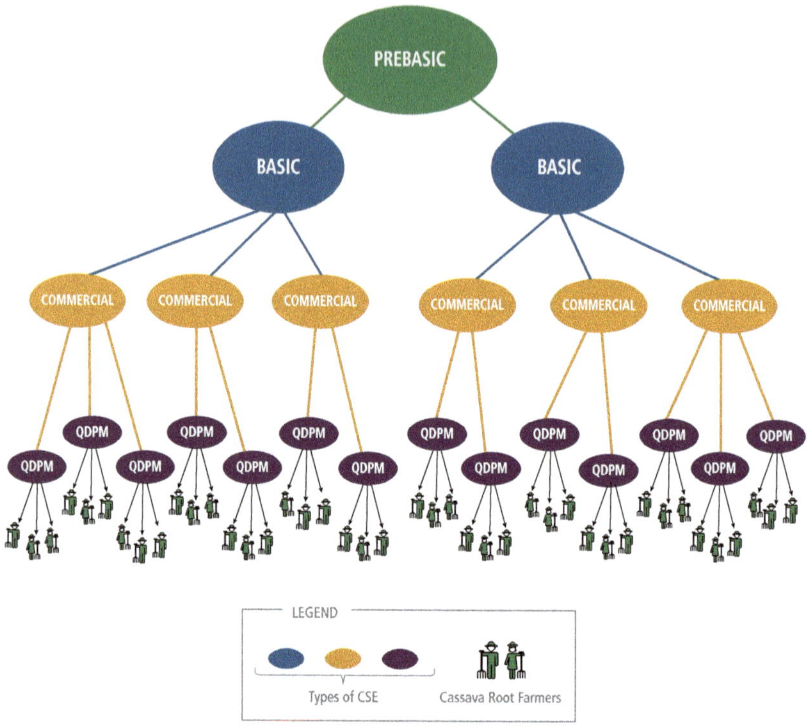

Figure 7 Hub-and-spoke model of the cassava seed system.

model, there were some CSEs higher up the chain (Basic and Commercial) that could identify their new customers easily. On the negative side, there were no advance sales contracts in place and production planning was difficult to meet anticipated demand one year down the road. In addition, the model for the cassava seed production included 'ratooning'; the practice of harvesting of stems to be sold while leaving roots in the ground to sprout new stems in another cycle. This enabled CSEs to yield three cycles of stems from one planting of cassava cutting seed. However, this also meant for all seed producers that they may not have annual repeated sales and will need to search for new customers on years 2 and 3. These downstream growers may harvest stems for two cycles before replanting, thus the producers need to plan adequate demand every three years instead of annually.

Establishing price and convincing farmers that they should pay for cassava seed was another challenge. There were two contributing factors to this. First, farmers traditionally have kept their own seeds for reuse in their farms, while others obtained the planting materials from friends and relatives. Second, there was a relatively common practice of NGOs and district governments distributing cassava seed free of charge to farmers with the goal to improve food security in the regions. While the free distribution of seeds was well-intentioned, its reach was limited, and the number of farmers targeted was constrained by project budgets. Given the transient nature of NGOs and fluctuating local budgets, one year of free distribution may be followed by another without free distribution. While these efforts met immediate needs for household food production, it conversely created a system of dependency, whereby, farmers expected to access quality planting material through free distribution. It effectively killed the potential for private sector market for purchasing clean, certified seed as farmers were unwilling to pay for materials they often received for free.

A third challenge in market access was the transportation of the cassava cuttings. Since cassava cuttings are perishable and bulky to transport, this generated significant transportation costs in some areas and logistics had to be carefully planned. The perishable nature of cassava stems is both a benefit to the market and a detriment. Because it cannot last long after being cut, the distance between seed producers and customers must be relatively short. This means that those who are far from seed growers may not be able to access the input. There is a positive side to the perishable nature of cassava cuttings: since cassava cuttings are not typically taken to market, but instead harvested and sold on the farm, being a CSE is attractive to women smallholders as they do not have to travel to market and can negotiate prices themselves. This provides them with better bargaining power and does not subject them to the cultural/physical barriers of having to bring their product to market. However, those who are close enough to access cassava seed are often close enough to build relationships with their supplier. Customers are able to communicate their needs or the difficulties they are facing with varieties. This builds a stronger overall market network to mitigate some of the demand and supply challenges listed above. Given these factors, there is a growing recognition of the importance of limiting and reducing market distortions and establishing a vibrant market for cassava seed.

There have been a number of activities that the project has facilitated to address challenges in market access. First, it has engaged with the public-sector extension agents and district and regional-level government representatives to ensure support and leverage their connections with other farmers and stakeholders in the regions. Second, it facilitated the organic development of farmer associations of CSEs. Third, it assisted with initial marketing and promotion activities which involved all stakeholders from individual CSEs, the associations and the government representatives. Now growers' associations assist with marketing and promotion across the levels and smooth areas of supply and demand shortfalls by connecting customers with other producers in the region. However, with high demand, many CSEs do not feel the need to participate in these associations (membership requires a fee, there is still mistrust surrounding associations given some poor experiences with cooperatives).

9.8 Decision-making

The experience working with smallholder farmers acting as commercially focussed entrepreneurs in the cassava seed system has yielded a number of learnings on their decision-making process and priorities.

Learning #1 - Farmers continually adjust their business models over time to maximize their profits.

As mentioned previously the business model for the CSEs was primarily focussed on cassava seed production, harvesting stems for multiple cycles while leaving roots in the ground. While income can be generated from the stems, farmers are still wanting to sell some roots for additional income. Some of these farmers opportunistically sell the roots when they see the price is high in the market. Others may plan ahead to use some of their field for roots and some for stems. The challenge is that the optimum time to sell roots (dry season) is offset from the stems (rainy season) so by selling one, you may have to forego the income from the other. Some CSEs have been creative in finding solutions either at short-term or longer-term basis. In the short term, some CSEs 'milk' their field selectively taking only a few roots from each plant to maintain the stem for harvest. This is resource-intensive and gets incremental root revenue. In the longer term, some CSEs divide their fields into two and have alternating cycles of root and stem for harvest. This requires larger land allocation for sufficient stem production. Others may opt to sell the entire field of stems at first rainy season (12 months after planting) and then roots at subsequent dry season (18 months after planting). This model foregoes the second stem harvest and therefore no income at 24 months after planting. These examples illustrate the complexity of the decision-making process of the farmers as well as their tendency to change business models based on expectation of price of roots compared to stems. It also shows that some farmers may be more opportunistic, choosing short-term options that may impact a longer-term income potential.

Learning #2 - Farmers base decisions on what they see and experience more than hypothetical plan.

There is a difference between CSEs who have been with the project for a few years compared with those who are in the first year. The experienced farmers have increased field size and seen profit in the cassava seed business. Those who are just entering the business are unsure as they may only see profitability after three years. In addition, the demonstration effect in the project is high. Those who have been profitable have nicer farms and have been visibly successful to their neighbours. This 'demonstration effect' leads to increasing numbers of farmers wanting to start a cassava seed business, with the recognition that the payoffs might be slow, but they are positive.

Similarly, a CSE's geographic distance to the CSE association (and thereby where the revolving loan fund was managed) was predictive of their desire to apply for a loan and contribute equity to the association. Those who were closer to the CSE association were more likely to put equity into the association and borrow from it. Smaller distances meant that CSEs could monitor and observe the fund and had higher confidence in the mechanism. While the fund was being established, those who were further away were more reluctant to put in money, and when the fund did not take off immediately, were more likely to pull their equity.

Learning #3 - Support services and output markets are strong determinants of farmer behaviour.

Beyond the immediate value chain for cassava seed, there are other factors/actors within the cassava market system that could influence it heavily in a way that would promote demand for quality seed. One of these factors is the presence of strong output markets, which to date, have not spurred demand as they have in other countries. In cassava-growing countries like Nigeria and Brazil, cassava processors demand high-quality cassava, and either source planting material for their out-grower network from quality sources or are seed dealers themselves. While there have been rumours of Chinese investment in Tanzanian cassava and demand for starch, these have not bolstered the seed system to date.

Related to this, government forces can limit the production of crops, despite possible output markets. For example, Tanzania has previously placed export bans on maize which prevents farmers from reaching higher income markets. These bans have been placed on grounds of food shortage in some parts of the country as well as a deliberate move to bring the inflation down. While these have not occurred for cassava, as a staple crop, there is a risk that the same could occur. Farmers are aware of these larger trends and their willingness to expand their operations is influenced by these actors.

10 Summary

Understanding the economics of smallholder farming is an important component in the search for sustainable intensification of farming systems. In economics, it is helpful to recognize that smallholder farmers are not one homogeneous group but rather can be grouped into segments that may make different decisions when faced with similar constraints. Segmentation can be based not only on farm size but also on other factors such as gender considerations, livestock holding, amount of revenue, share of crop sold, share of income from agriculture and reliance on family labour. A gendered analysis of these elements leads to the recognition that some smallholders may be categorized as subsisting while others may be more commercialized in nature. Subsequently designing for these factors primes a project for effective implementation and provides a framework for systematic change for women and men smallholder farmers to succeed.

This chapter outlined the key economic factors that impact smallholder farmers and the constraints faced by smallholders that tend to keep them marginalized within broader markets. Land, labour, capital and inputs are economic factors that influence smallholder farmer profitability by directly impacting the production costs. Access to markets influences the farmer's ability to effectively sell their products and affects both the distribution costs and the revenue. Each of these economic factors has an inherently gendered dimension and can be combined as inter-related constraints in the farmers mind as decisions about what crops to produce, how to optimize production and where to sell it.

There is not a 'silver bullet' solution that fits all situations that smallholder farmers face. Solutions that alleviate the challenges that smallholder farmer face explicitly in these economic conditions are context-specific and need to have robust analysis and planning based on farmer experience and sound market information. Some of these solutions may be found in several areas:

1. Policy-related solutions such as those related to banking and microfinance (addressing access to credit) or those related to infrastructure projects (addressing access to market).
2. Technology-based solutions such as those using mobile/internet technologies that provide better access to markets, technology to reduce labour and more information for farmers to make decisions.
3. Temporary incentives to access new inputs such as targeted one-time grants or cost-sharing arrangements implemented with incentive to keep purchasing on an ongoing basis.
4. Structural changes that improve access to information, finance or building capacity. Examples include demonstration plots (access to information for inputs), revolving loan fund (access to finance) and training of extension agents to help farmers with inputs and manage their farms (building capacity).

Solutions must also keep in mind the complex context that the smallholder farmer operates within and the corresponding decision-making process factoring in the multiple constraints that, when addressed sufficiently, will lead to sustainable ways to overcome them.

11 Where to look for further information

Agricultural Development Book

Seminal work on agricultural development and economic transformation based on small commercial farmers.

- Mellor, J. (2017). *Agricultural Development and Economic Transformation*. Cham: Springer International Publishing.

Segmentation and Financial Inclusion

- Consultative Group to Assist the Poor (CGAP): https://www.cgap.org/.

Decision-Making Resources

- Guide to developing a decision matrix (Grain & Graze Australia): http://www.grainandgraze3.com.au/resources/Guide%20to%20developing%20a%20decision%20matrix%20(V2).pdf.
- Risk and Decision Game (Grain & Graze Australia): http://www.grainandgraze3.com.au/cb_pages/risk_and_decision_game.php.

12 References

Anderson, J., Collins, M. and Musiime, D. (2016). *National Survey and Segmentation of Smallholder Households in Tanzania Understanding Their Demand for Financial, Agricultural, and Digital Solutions*. Consultative Group to Assist the Poor (CGAP), Washington, D.C.

Anderson, J., Karuppusamy, R., Neumann, P. E., Miller, H. and Tamara, R. (2019). *Smallholder Households: Distinct Segments, Different Needs*. Consultative Group to Assist the Poor (CGAP), Washington, D.C.

Boniphace, N. S., Fengying, N. and Chen, F. (2015). An analysis of smallholder farmers' socio-economic determinants for inputs use: A case of major rice producing regions in Tanzania, *Russian Journal of Agricultural and Socio-Economic Sciences* 38(2), 41–55.

CHRAGG (2016). *Shadow Report Relating to the 7th and 8th Tanzania Report on the Convention on the Elimination of All Forms of Discrimination Against Women (CEDAW)*. Commission for Human Rights and Good Governance (CHRAGG), Dar es Salaam, Tanzania.

FAO (2002). *Land Tenure and Rural Development. FAO Land Tenure Studies 3*. Rural Development Division, FAO, Rome.

FAO (2011). *The Role of Women in Agriculture*. ESA Working Paper No. 11-12. Agricultural Development Economics Division, FAO, Rome.

FAO (2014). *Understanding Smallholder Farmer Attitudes to Commercialization, The Case of Maize in Kenya*, Woolverton, A. and Neven, D. (Eds). FAO, Rome.

FAO (2018). *Proposed International Definition of Small-Scale Food Producers for Monitoring the Sustainable Development Goal Indicators 2.3.1 and 2.3.2*. Office of the Chief Statistician and the Statistics Division, FAO, Rome.

Gourlay, S., Kilic, T. and Lobell, D. (2017). *Could the Debate Be Over? Errors in Farmer-Reported Production and Their Implications for the Inverse Scale-Productivity Relationship in Uganda*. Policy Research Working Paper 8192. World Bank Group, Development Economics - Development Data Group. Available at: http://econ.worldbank.org.

Hepelwa, A. S., Selejio, O. and Mduma, J. K. (2013). *The Voucher System and the Agricultural Production in Tanzania: Is the Model Adopted Effective? Evidence from the Panel Data Analysis*. Environment for Development EfD. Available at: https://efdinitiative.org.

Isaga, N. (2018). Access to bank credit by smallholder farmers in Tanzania: a case study. *Afrika Focus* 31(1), 243–244.

Mellor, J. (2017). *Agricultural Development and Economic Transformation*. Springer International Publishing, Cham.

Mercy Corps (2019). *AFA Case Study: Digital Pathways for Youth in Agriculture*. Mercy Corps. Available at: https://mercycorpsagrifin.org/wp-content/uploads/2018/12/230118_afa-youth-final-vF-compressed.pdf.

Murphy, S. (2012). *Changing Perspectives: Small-Scale Farmers, Markets and Globalization* (revised edition). IIED/Hivos, London/The Hague.

Ndyetabula, I. L., Merumba, S. M., Jeremiah, S. C., Kasele, S., Mkamilo, G. S., Kagimbo, F. M. and Legg, J. P. (2016). Analysis of interactions Between cassava brown streak disease symptom types facilitates the determination of varietal responses and yield losses, *Plant Disease* 100(7), 1388–1396. Available at: http://dx.doi.org/10.1094/PDIS-11-15-1274-RE.

Peck, R., Anderson, C. and Anderson, J. (2013). *Segmentation of Smallholder Households: Meeting the Range of Financial Needs in Agricultural Families*. Focus Note 85. Consultative Group to Assist the Poor (CGAP), Washington, D.C.

Rapsomanikis, G. (2015). *The Economic Lives of Smallholder Farmers: An Analysis Based on Household Data from Nine Countries*. Food and Agriculture Organization of the United Nations, Rome.

Shirima, R. R., Maeda, D. G., Kanju, E. E., Tumwegamire, S., Ceasar, G., Mushi, E., Sichalwe, C., Mtunda, K., Mkamilo, G. and Legg, J. P. (2019). Assessing the degeneration of cassava under high-virus inoculum conditions in coastal Tanzania, *Plant Disease* 103(10), 2652–2664.

SICGroup (2017). *Responsible Agriculture Finance for Smallholder Farmers in Tanzania and Uganda*. Report prepared for The MasterCard Foundation. Social Innovation Consulting Group. Available at: https://www.raflearning.org/post/responsible-agriculture-finance-for-smallholder-farmers-tanzania-and-uganda.

Sipangule, K. (2017). Agribusinesses, smallholder tenure security and plot-level investments: evidence from Rural Tanzania, *African Development Review* 29(S2) (Suppl.2), 179–197.

Tanzania Local Law (2002). Local Customary Law (Declaration) (No. 4) Order, Government Notice (GN) 436/1963, Schedule 1, Laws on Guardianship [Sheria za Ulinzi]. In: *Judicature and Application of Laws Act*, TANZ. LAWS SUBSIDIARY LEGIS. [*CAP 358 R.E. 2002*], Tanzania.

Chapter 5

The economics of soil health

*Maria Bowman, USDA-ERS, USA**

1 Introduction

Much of the current and previous research on the economics of soil conservation and soil health has focused on the farmer's adoption decision as it relates to discrete soil health practices, such as no-till or reduced tillage, cover cropping or nutrient management (see USDA Natural Resources Conservation Service (2017) for a list of practices considered a part of a soil health management system). This research is data-intensive, as it requires data that link on-farm soil health variables or changes in soil health or environmental benefits to changes in practices *and environmental and economic outcomes*. In addition, economists have often focused on evaluating one practice or programme at a time (or one outcome variable, such as yield or profit) in order to more easily identify the effect or change they care about. However, this historical focus on individual practices in research and in public policy design means that there is less economic research that examines whether soil health practices are substitutes or complements for other soil health practices in terms of the production of benefits for agricultural production, or of environmental benefits.

* The views expressed are those of the author and should not be attributed to the Economic Research Service or USDA.

http://dx.doi.org/10.19103/AS.2017.0033.06
Published by Burleigh Dodds Science Publishing Limited, 2018.

Economists' contributions to research on the economics of soil health go beyond efforts to describe the current patterns of adoption and quantify the costs and benefits of soil health and soil health practices to farmers and society. Using econometric techniques, socioeconomic surveys, spatially explicit modelling and behavioural experiments, economists are researching the patterns of adoption of soil health practices, the barriers to adoption of soil health practices, the role of cost-share or incentive programmes in farmer decisions to adopt soil health practices and the cost-effectiveness and optimal design of incentive programmes.

This chapter looks at basic patterns of adoption of a few key soil health practices with a focus on the United States, followed by a discussion of factors that affect the farmer's adoption decision about soil health practices. The bulk of the chapter discusses several active areas of research that relate to the economics of soil health and provide context for why these research areas are important for public policy.

2 Use of key soil health practices by farmers in the USA

In the United States, there are several farmer surveys administered by the U.S. Department of Agriculture (USDA) that are nationally representative and collect information about soil health practices including tillage practices, cover cropping, some nutrient management practices, soil health testing and rotational grazing. These data allow us to track farmer adoption rates in space and time. Data from the most recent Census of Agriculture, conducted in 2012, suggest that more than a third of farmland area is engaged in no-till agriculture, while less than 3% of cropland area is cover-cropped. However, approximately 9% of the *farms* with cropland report using cover crops on some portion of their farm. Although these are the national rates for no-till and cover crop adoption, the rates of adoption of these practices vary by region, crop grown and other variables (Wade et al., 2015). Wade et al. (2015) use several years of data from the USDA Agricultural Resource Management Survey (ARMS) to show how adoption rates of tillage practices and cover cropping vary over time, as well as regionally. The authors found that no-till/strip-till and cover cropping are used more frequently in the southern and eastern regions of the United States, and that no-till/strip-till use is the highest on soybean fields (between 40 and 45% of the cropland planted to soybeans used no-till in the 2006 and 2012 versions of the survey) when compared to cropland planted to corn, wheat and cotton. Wade et al. also report that 64% of the cropland planted to cotton in 2007 and 31% of the cropland planted to corn in 2010 reported splitting applications of nitrogen fertilizer so that parts of the crop's nitrogen requirements were applied after

Table 1 Adoption rates in the United States for several key soil health practices

		Adoption rate	Year	Data source
Cover cropping	Wade et al. (2015)	<2% of corn, soybean, wheat, and cotton planted area	United States, 2010–11	USDA Agricultural Resource Management Survey
	Bowman et al. (2016)	8.6% of farms with cropland and 2.9% of cropland	United States, 2012	USDA Census of Agriculture
No-till/strip-till	Bowman et al. (2016)	34.6% of cropland in no-till (for which tillage practices are reported)	United States, 2012	USDA Census of Agriculture
	Wade et al. (2015)	39% of corn, soybean, wheat, and cotton planted area in no-till/ strip-till	United States, 2010–11	USDA Agricultural Resource Management Survey
Combined nutrient management practices[1]	Wade et al. (2015)	24% of cotton planted area 6% of corn planted area	United States; 2007 (cotton) and 2010 (corn)	USDA Agricultural Resource Management Survey
Testing for soil organic matter	Bowman et al. (2016)	11% of oat fields and 15.8% of cotton fields surveyed tested at least once in last ten years	United States, 2015	USDA Agricultural Resource Management Survey
Rotational or management-intensive grazing		Approximately 23% of farms with pastureland[2]	United States, 2012	USDA Census of Agriculture

[1] No fall application of N, some N application after planting, N application at rates below a 'benchmark' and fertilizer incorporation or injection below the soil surface.
[2] 288 719 farms/1 239 085 farms with pastureland (excluding woodland pastured).

planting. In addition, 24% of the cropland planted to cotton and 6% of the cropland planted to corn reported combining split nitrogen application with no fall application, some nitrogen application after planting, applying less than the recommended amount of nitrogen and fertilizer incorporation or injection. Table 1 summarizes the most recent national adoption rates for soil health practices reported in USDA datasets.

Adoption rates of soil health practices vary regionally due to a number of factors (Wade et al. 2015). These include inherent differences in agronomic factors such as soil types and climatic factors, demographic and social factors and economic factors. State and federal incentive programmes, policies and regulations also impact adoption rates, as do regional differences in farmer outreach and extension activities. Figure 1 shows dramatic regional differences in cover crop adoption patterns in the United States.

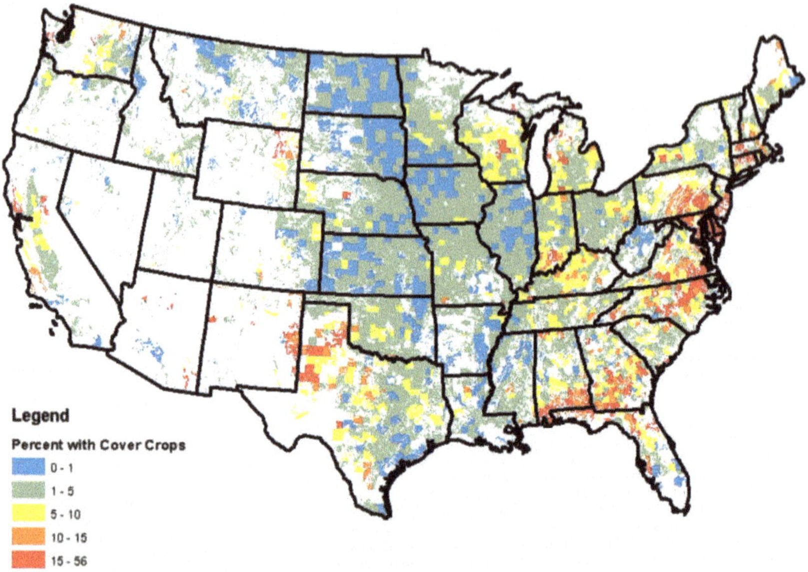

Figure 1 Share of harvested cropland area with cover crops by county in the United States as reported in the 2012 USDA Census of Agriculture. Source: Bowman et al. (2016), using data from the 2012 USDA Census of Agriculture.

3 Costs and benefits of soil health practices

When farmers make decisions about adopting soil health practices, they are usually weighing whether the benefits they receive from adopting a practice or suite of practices exceed the costs of implementing the practice(s), and how these costs and benefits might compare to other types of production systems or practices over time. From an economic perspective, these types of costs and benefits that accrue to or are borne by the farmer are referred to as *private* costs and benefits. Soil health practices can also have benefits for society, or *public benefits*. Table 2 summarizes the potential private costs and benefits associated with cover cropping, a key soil health practice. Snapp et al. (2005) summarize some of the key private costs and benefits associated with cover cropping in Michigan. In general, private benefits of adopting one or more soil health practices might be improved crop yields, reduced agricultural input costs due to reduced need for synthetic fertilizer or other inputs, greater resilience of crops to extreme weather (droughts and floods) as a result of improved soil water-holding capacity, and an opportunity to engage in environmental stewardship. Private costs could include investments in time or resources as the farmer learns about a new practice, increased labour associated with monitoring or managing (some studies cite this as a benefit - that is, decreased labour requirements associated with using soil health practices relative to

Table 2 Potential private costs and benefits of planting a cover crop

Costs	Benefits
• Investment in learning about new practice • Cover crop seed and other costs associated with establishment • Increased herbicide use, if chemically terminated • Increased labour associated with seeding, managing, and terminating cover crop • Decreased yield (yield penalty) in cash crop • Opportunity cost of not planting cash crop, if there is a cash crop option • Potential loss of flexibility due to timing of planting and termination of cover crop with harvest and planting of cash crops and/or poor establishment of cash crop • Difficult to estimate N benefit and timing of N release • Competition for water in water-scarce regions	• Improved yields (yield boost) • Weed or pest suppression can result in reduced input costs for herbicides or pesticides • Provides nutrients (e.g. N from legumes) which can reduce need for fertilizer inputs • Increased water use efficiency might reduce irrigation costs or improve yields • Greater resilience to extreme weather events such as droughts and floods can mitigate yield losses • Improved nutrient cycling • Reduced erosion and/or improved soil stability/soil structure • Pride in farm and/or in environmental stewardship • Grazing (forage) benefit (if grazed)

Note: Some items listed could fall in either column depending upon the direction of the effect on yield, planting time, labour requirements or other variables.

conventional management practices) and increased costs of certain other inputs (such as cover crop seeds or additional herbicides).

One body of literature related to soil health practice adoption aims to simply estimate when and where it makes economic sense to the farmer to adopt soil health or conservation practices, and what the production outcomes or expected financial returns are to the farmer compared to other practices or combinations of practices. These types of studies are sometimes based on the data from field/farmer trials, models, or farmer surveys. For example, Karlen et al. (2013a,b) evaluate the impacts of different types of tillage over 30 years in Iowa on yield and soil fertility, as well as the production costs associated with each. Corn and soybean rotations boosted corn yield by 17% when compared to continuous corn, and although tillage produced the highest yields, no-till had the highest net returns over the study period due to reduced costs. They also confirm that mouldboard ploughing over the long run has a significant and negative effect on soil quality. Al-Kaisi et al. (2016) also found that the net return is higher for no-till corn and soybean rotations in a study in Iowa due to reduced input costs and comparable yields, and Cook and Trlica (2016) show that yield penalties associated with no-till over the long term in Southern Illinois disappear with adequate NPK.

Box 1 Risk and farmers' decision-making

Farming is an inherently risky endeavour, and farmers face different types of risk. Within a growing season, these include price risks (i.e. the risk that input prices or commodity prices will change) and yield risks, or production uncertainty (i.e. the risk that a farmer's crop will not produce the expected yield at the expected time), among others (Chavas et al., 2010). There are also risks that farmers must consider over a longer time horizon, such as the risk associated with trying a new technology.

Farmers have many tools at their disposal to manage risks. These include using credit markets to borrow at times of low business cash flow, participating in government insurance programmes, engaging in contracts that guarantee a set purchase price or volume for inputs or for a final agricultural product and diversifying input or output mixes. Finally, farmers can use management practices that may help to mitigate risks.

Adopting soil health practices and systems has the potential to increase or decrease risk. Adopting soil health practices might increase the yield risk due to pest or weed pressure, particularly in the short term, if a farmer is trying out a practice for the first time. On the other hand, soil health practices may decrease risks by making a farmer more resilient to droughts or extreme weather, for example.

Hanson et al. (1993) and Lichtenberg et al. (1994) evaluated the profitability of cover crops for no-till corn systems in the Mid-Atlantic region. Hanson et al. (1993) found that no-till corn after a hairy vetch cover crop was more profitable than either after winter fallow or a winter wheat cover crop in the Piedmont and Coastal Plain regions of Maryland. Lichtenberg et al. also found hairy vetch to be more profitable than fallow and cover crop alternatives, and that this was primarily due to a yield advantage for hairy vetch. Watkins et al. (2004) found that no-till may be more profitable than conventional tillage in rice systems in Arkansas in some years but not in others.

These types of studies highlight an important point: though agronomic studies often focus on what practices have the highest yield, high-yielding practices are not inherently the most profitable practices. Where this is the case, it might be due to reduced input use – and, in turn, input costs – associated with the lower yielding practice relative to conventional systems that may not engage in soil-health-related practices. A farmer may also care about yield variability in addition to just average yield; so to the extent that soil health practices impact average yields and yield variability, this will also affect farmer decision-making about soil health practices (see Box 1 'Risk and farmers' decision making').

This potential benefit of soil health practices, risk mitigation, is an increasing focus of the agronomic and economic literatures. The risk of drought is perhaps the most important weather/climate-related risk a non-irrigated farmer faces, and the economic literature suggests that soil health practices become a tool that farmers use to mitigate and adapt to drought risk. Between 1948 and 2010, 38% of the crop insurance indemnity payments, at a value of $32.5 billion, were due to farm production losses from drought (followed by excess moisture, at 24% or $20.5 billion) (Wallander et al., 2013). Counterintuitively, when controlling for recent indemnity and ad hoc disaster payments, Schoengold et al. (2015) found that the number of recent flood years increased the use of no-till and that the number of recent drought years decreased the use of no-till. They also found that counties with high disaster and crop insurance indemnity payments were more likely to use no-till and less likely to use other types of conservation tillage. Their results suggest that there may be complex interactions between federal crop insurance and disaster payment programmes and farmer decision-making, and that future research might consider how such programmes impact farmer behaviour and may incentivize or disincentivize farmers to engage in on-farm risk management through the adoption of practices (such as conservation tillage).

A slightly different collection of the literature looks at the role of specific factors in influencing a farmer decision to adopt a practice, often while controlling for other factors known to influence farmer decision-making (such as input prices, crop prices, location, agri-environmental variables, government programme participation, farmer characteristics, etc.). We already discussed how location-specific biophysical variables might affect farmer adoption decisions and drive differences in regional adoption rates. But, what about other variables like farm/field size and farmer characteristics, and economic variables like prices? Prokopy et al. (2008) and Baumgart-Getz et al. (2012) conducted a review and a meta-analysis (respectively) of the published literature in order to determine the factors affecting the adoption of best management practices (BMPs) by farmers. Some of the factors that have a positive relationship with BMP adoption across the studies reviewed in Prokopy et al. (2008) and Baumgart-Getz et al. (2012) include capital, farm size, specialization in farming, access to information and environmental attitudes/awareness.

Characterizing patterns of practice use over time and space is also important in order to understand public and private benefits, as well as farmer decision-making. Understanding these dynamics is especially important for practices that farmers can make choices about every year or planting season. For example, the adoption of many practices, such as no-till and cover crops, can occur in some fields but not others, or in one year but not the next. This may have something to do with changes under annual weather conditions, or with crop choice and crop residue management. There may also be on-farm or off-farm spillover effects. Farmers that adopt a practice on a field may be more likely to adopt it on others if, for example, they learn how to successfully implement the practice on a small portion of the farm and later expand, or if they have to purchase new farm machinery (capital) in order to adopt a practice. Off the farm, farmers share knowledge and information within their social and professional networks about practices and about the economics of practice adoption (Foster and Rosenzweig, 1995). If a practice is observed to be profitable in a region or if the learning costs are reduced by transmission of information, other farmers may be more likely to adopt, and social norms and attitudes about particular practices may also change as more farmers adopt. Sunding and Zilberman (2001) discuss different models of how patterns of technology adoption can be described over space/time, such as the diffusion and threshold models of adoption.

In the case of soil health and conservation technologies, social and professional groups in the United States provide important opportunities for farmers and researchers to network with peers and learn about technologies through outlets, such as publications, radio broadcasts and conferences. For example, the National No-Till Conference has been occurring annually for 25 years, and the newly established National Conference on Cover Crops and Soil Health is organized by the Soil and Water Conservation Society (Meitner, 1999; No-Till Farmer, 2017; Soil and Water Conservation Society, 2017).

4 Case studies in soil health: strengths and limitations

Case studies are frequently employed by government agencies and extension agents to illustrate the costs and benefits of adopting a particular practice. Table 3 presents a number of soil health case studies that are publicly available to farmers and researchers. Another publicly available source of case studies is the 'Profiles in Soil Health' published by USDA Natural Resources Conservation Service (NRCS) (Natural Resources Conservation Service, 2016). These compelling discussions with farmers highlight the cases where managing for soil health resulted in large soil health gains and was profitable for the operations.

Case studies are important in disseminating real-world experiences with soil health practices or combinations of practices. They are also written so as to be accessible to farmers and practitioners, and to focus on things that farmers care about (rather than academics). Another strength of case studies is that they are often inexpensive to research and write, and can provide early intuition about the economic variables that affect farmer decision-making. But they have the disadvantage of being a single data point (or few data points, if farmers engage in small on-farm trials), which makes it difficult to draw statistical conclusions or generalize the results to other farmers, regions or policy situations. Methods are also not commonly standardized across case studies, which means that the quantification of costs and benefits and any

Table 3 Selected economic case studies in soil health

Case study author(s)*	Farmer/Farm and location	Practices adopted	Challenges/Costs	Benefits and economic data points
Susan Tallman, National Center for Appropriate Technology	Gabe Brown (North Dakota)	• No-till • Diverse cover crops • Intensive rotational grazing		• SOM increased from 2 to 4% • Improved water infiltration • Reduced fertilizer and herbicide use • Average yield much higher than county average
Madalene Ransom, USDA NRCS	Russell Hedrick (North Carolina)	Cover crops added to no-till corn and soybeans	• Learning costs in year 1 • High costs for seed and herbicides	• 927 kg/ha higher yield for corn (9%) • 1,201 kg/ha higher yield for soybeans (35%)
Susan Tallman, National Center for Appropriate Technology	Bauer Farm (North Dakota)	Cover crops and grazing (for corn) added to no-till wheat and corn	• Adapting rotations to no-till	• No difference in wheat yields between CC field and control • Corn yields for CC + grazing were 6,178 kg/ha (20% higher) compared to 5,127 kg/ha in control
Susan Tallman, National Center for Appropriate Technology	Richter Farm (North Dakota)	Grazing + cover crops in forage based system, followed by corn		• Corn with cover crop had a net return of 153.85 USD/ha vs. no cover crop of 34.03 USD/ha (452% higher net return)
Natural Resources Conservation Service, New York	Dave Magos (New York)	Cover crops and no-till in corn silage and alfalfa rotation		• Reduced soil erosion, and pesticide and fertilizer use • 5% increase in corn silage yields • Net return on cover crops is 153 USD/ha
Conservation Cropping Systems Initiative	Rulon Enterprises (Indiana)	Cover crops added to no-till corn and soybeans	Working with Purdue and NRCS in initial trials may have helped with learning costs	• Yield over county yield higher during drought years • 439 kg/ha yield boost for corn and 133 kg/ha for soybeans over three years • Net benefit of 200 USD/ha for cover crops inc. CSP payment, expected yield boost in drought year, value on reduced soil erosion
Lauren Cartwright, MO State Economist, NRCS	(Anonymous)(Missouri)	Cover crops added to no-till corn and soybeans		• Yield increase for corn but not soybeans • Short term annual net cost • Herbicide reduction for soybeans • N credit for corn

* Full case study reference and link to online version (if available) provided in References.

estimates of net costs and benefits are highly dependent on the assumptions used and information collected as part of the case. Developing standards and guidelines for future case studies will be essential for comparability across case studies in soil health, though a meta-analysis may also prove useful. See Knight (2016) for an example of guidelines for developing case studies.

Case studies are also prone to issues of bias. In particular, farmers that manage for soil health and adopt soil health practices (as well as farmers that are willing to participate in case studies) are often not representative of other farmers. They may be more 'progressive' in their farming practices, be more accepting of risk or uncertainty, take more pride in environmental stewardship or be better or more holistic land managers than their peers (Dumanski et al., 2014). Early adopters are also typically those who benefit the most from adopting soil health practices (see Wu and Babcock (1998)). As a result, economic estimates of the costs and benefits of soil health practices based on these farmers are likely to be biased. In addition to the selection bias associated with the adoption of practices, case studies may also suffer from a form of publication bias. For example, because the purpose of a case study might be to provide an example of a success with a practice so that other farmers can overcome their misgivings about adopting a new technology or practice, it may be more difficult to find published examples of case studies where the outcome was not as successful.

Much of this research is conducted with an end goal of providing useful information to producers about the costs, benefits and profitability of soil health practices. To this end, a number of organizations and research groups have developed online decision tools or spreadsheet models that allow producers to input parameters that describe their system and receive outputs that reflect the impacts of different choices or practices on costs, yields or profitability.

Box 2 Economic decision tools for soil health practices in the United States

Cornell: Cover Crops for Vegetable Growers
http://covercrops.cals.cornell.edu/decision-tool.php
NRCS: Cover Crops Economics Tool
https://www.nrcs.usda.gov/wps/portal/nrcs/detailfull/il/soils/health/
Midwest Cover Crops Council: Selector Tool for Cover Crops
http://mccc.msu.edu/selector-tool/
Ohio State: Soil Organic Matter Calculator
https://southcenters.osu.edu/soil-water-bioenergy/extention/som-soil-organic-matter-calculator
Kansas State: Cover Crop Cost-Return Budget
http://www.agmanager.info/production-economics/farm-management-guides/livestock-budgets/ksu-cover-crop-cost-return-budget
Useful to Useable: Corn Split-N tool
http://mrcc.isws.illinois.edu/U2U/splitn/

Box 2 'Economic decision tools for soil health practices', contains a short list of some publicly available decision tools related to soil health economics and management.

5 Public benefits of soil health and soil health management practices

In addition to the costs and benefits of changes in soil health and soil health practices that are relevant to the farmer, soil health can have important public benefits. These can include reduced greenhouse gas emissions from agriculture, increased carbon storage or sequestration, improved water quality through reduced nutrient and sediment pollution and impacts on water quantity/availability (Dale and Polaskly, 2007; Hansen and Ribaudo, 2008; Hatfield et al., 2001; Lal, 2004; Ribaudo et al., 2010; Swinton et al., 2007). Quantifying and valuing these public benefits is difficult, because they accrue at different scales and are difficult to measure. Nonetheless, there is theory and evidence to support the fact that these benefits could be significant, and valuing them is important when considering the costs and benefits of public policy (see Robinson et al. (2012) for a framework for valuing the natural capital of soil). In particular, the relationship between soil health and water quantity/quality has the potential to generate significant public benefits due to improved infiltration rates and increased soil water storage capacity (Hudson, 1994; Rawls et al., 2003: Kemper et al., 2011). This may contribute to the following:

- Reduced severity of flooding during extreme precipitation events.
- Reduced sedimentation in waterways and reduced costs associated with dredging public reservoirs.
- Reduced demand for irrigation water, which could have a positive impact on groundwater and surface water supply, as well as demand for water deliveries from federal or state programmes.
- Improved resilience to drought, which could reduce federal indemnity payments for yield losses incurred during drought years.
- Reduced nutrient pollution at local, regional and watershed scales due to reduced runoff from farm fields.

In the early 2000s, USDA began the Conservation Effects Assessment Project (CEAP) with the aim of understanding the environmental benefits of USDA conservation programmes (Mausbach and Dedrick, 2004). As part of this research, ARS completed several 'benchmark watershed assessments', which were designed specifically to look at the impact of agricultural conservation practices on water quality at the watershed scale. Tomer and Locke (2011) summarize the results of the assessments in the CEAP benchmark watersheds.

Models using results from the project predict that a number of conservation practices that are also soil health practices – such as conservation cover, cover crops, nutrient management, no-till and pasture rotation – will contribute to benefits such as reductions in sediment concentration and nitrate and phosphorus leaching and loading at multiple scales. However, they also discuss the limitations of the project in terms of available data and ability to link changes in practices to water quality outcomes. In particular, they note that monitoring water quality alone may not allow researchers to detect the specific effects of practices given climatic variation, monitoring issues associated with non-point sources such as agricultural fields and farms, and time lags from practice implementation to water quality effect detection.

In a 2016 report, USDA NRCS estimated that all voluntary conservation practices (including structural, nutrient management, cover crops and others) in the Western Lake Erie Basin (as of 2012) reduce annual sediment losses by 81%, nitrogen by 36% and phosphorus by 75% relative to a baseline of no conservation practices (U.S. Department of Agriculture, 2016). Though not all these practices were soil-health-related, these results illustrate the magnitude of the potential public benefits that can be achieved through conservation practice adoption.

In 2012, the Australian government sponsored a report summarizing the relationship between soil health management practices and ecosystem service provision (Cork et al., 2012). Some key findings of this report were that soil health management practices have slowed the loss of soil carbon and that this has contributed to public and private benefits; that maintaining high levels of residue cover (50–70%) has reduced wind and water erosion; and that improved management has reduced dust storms, which have significant private and public costs. See Robinson et al. (2012), Bennett et al. (2010) and Huguenin et al. (2006) for additional resources that discuss on valuing the public benefits (ecosystem services) associated with soil health and soil health management practices in different regions.

6 Barriers to adoption of soil health practices

Although there are public and private benefits to the adoption of soil health practices and management systems, they can also be costly to implement. A straightforward reason that farmers might not adopt soil health practices is that they may not be profitable on a particular farm at a particular point in time, or the farmers do not anticipate that they will be profitable in the long run. This could be due to a number of factors already mentioned in this chapter, including fixed or variable costs associated with soil health practices. Important among these may be investments in capital, such as new planters, or investments in human capital required for a farmer to learn how to implement a new practice

(Arbuckle and Ferrell, 2012; Carlisle, 2016). Risk or perceived risk of adopting soil health practices also influences farmer decision making (see the Box 1 'Risk and farmers' decision making'). Arbuckle and Roesch-McNally (2015) found that farmers who do not adopt cover crops often perceive higher levels of risk associated with the practice. The authors suggest that this highlights a need for improved communication with farmers about both the potential benefits and the risks of the practice, as well as a clear need for technical assistance and education.

There are also behavioural or informational barriers to the adoption of soil health practices and systems. These include social barriers to adoption, such as when practices are not viewed as compatible with current farming practices, or when low adoption rates influence attitudes towards a practice within a social network (Rogers, 2003; Reimer et al., 2012). Lack of information about the costs and benefits of soil health practices seems to be a significant barrier. Dunn et al. (2016) use a national survey of cover crop users to look at patterns in self-funding vs. programme participation, as well as perceived barriers to adoption and factors affecting continuous use of cover crops. The authors find that many see the high cost of cover crops, combined with uncertain benefits, to be a major barrier - and that those who use cover crops continuously are comfortable with engaging in a long learning process by trial and error about what works best on their farm. While trial and error may be an important tool for farmers, it implies that farmers have uncertainty and limited information about how to implement cover crops profitably on their operation.

Another important barrier in this category relates to the time lags in benefits or profitability associated with practices. Though the costs of implementing soil health practices (such as investments in human or physical capital and cover crop seed) are borne in the present, the farmer may not immediately reap the benefits of changing their practices. Finally, institutional and structural barriers can also impact adoption, such as transaction costs associated with participating in government programmes and land tenure issues that affect short- and long-term decision-making of land owners and operators. Carlisle (2016) provides an excellent narrative review of the literature on the barriers to adoption of soil health practices.

7 Evaluating the role of federal and state regulations, policies and incentive programmes

In part because of the public benefits associated with soil health and soil conservation, it may make sense to incentivize farmers to adopt soil health practices when they might not adopt them on their own. There are many United States policies that target farm management practices related to soil health. Several types of regulations directly affect farmer decision-making about soil health practices. Perhaps the best example of a United States regulation that affects soil health is 'conservation compliance', or provisions that require farmers to implement soil conservation practices on highly erodible cropland in order to retain eligibility for federal farm payment programmes such as conservation payments, crop insurance, disaster payments and loan programmes. Conservation compliance provisions were implemented as part of the 1985 Farm Act. During the years following the implementation of conservation compliance, erosion declined sharply on highly erodible cropland with about 25% of the total erosion reductions occurring on highly erodible cropland (Claassen et al., 2004). Because erosion also declined on the land that was not highly erodible or subject to compliance, it is difficult to evaluate the role of other factors affecting soil erosion during the same time period (see Box 3 'The challenge of additionality'). However, Claassen et al. (2017) showed that conservation compliance contributed to decreasing water erosion on cropland subject to compliance.

Voluntary national programmes can also induce farmers to engage in soil health management practices, for example, by providing a cost-share or financial incentive to a farmer to adopt a specific practice for a defined length of time. In the United States, two programmes that provide financial incentives to farmers to engage in soil health practices or 'enhancements' (as described because they take conservation practices to a higher level of stewardship) are the Environmental Quality Incentives Program (EQIP) and the Conservation Stewardship Program (CSP). These programmes are managed by the NRCS of the USDA. For both programmes, farmers engage in a contract with NRCS for specific practices/enhancements on specific acres over a defined time period (e.g. many EQIP no-till contracts are three-year contracts). The most popular EQIP practices for soil health include nutrient management, cover cropping, no-till, reduced tillage, conservation crop rotations, mulching and seasonal residue management. Between 2009 and 2013, NRCS increased the land area enrolled in EQIP for cover crops, as well as the total expenditures for the practice. At the same time, there was a decrease in the amount of land receiving payments and in total no-till payments. This may be because no-till has become more widely adopted, which has presented opportunities for the programme to focus on other practices that promote soil health but may have a higher cost to the farmer, such as cover crops (Bowman et al., 2016). Table 4 lists some enhancements that are funded through CSP. The number of hectares receiving CSP payments for at least one soil health practice or enhancement grew from under 3 million ha to more than 12 million ha between 2010 and 2014 (Bowman et al., 2016). Zimmerman and Britz (2016) evaluate the participation of farms in the EU in agri-environmental measures or voluntary conservation programmes that are implemented as part of the EU's Common Agricultural Policy.

Box 3 The challenge of additionality

Economists and social scientists face a research challenge: many farmers implement soil health or soil conservation practices even in the absence of regulations or incentive programmes, which makes it difficult to quantify how the farmer behaviour changed in response to the regulation or programme. The additional effect of a policy, above and beyond baseline adoption, is known as 'additionality'. Research on additionality in conservation programmes is an important tool for evaluating how well they work, and whether they are cost-effective (see Claassen et al. (2013)).

In the United States, state-level incentives are another important source of financial assistance for soil health practices. For example, the state of Maryland's Agricultural Water Quality Cost-Share Program provides cost-share payments that cover up to 87.5% of the costs of installing a number of conservation practices, including cover crops (Maryland Agricultural Water Quality Cost-Share Program, 2016). In 2004, the programme paid farmers a minimum of 49.40 USD/ha to plant a cover crop. In that year, Maryland enrolled only 46,750 ha - but after raising the minimum payment to 111.20 USD/ha, more than 192,452 ha were enrolled in 2015 (Bowman et al., 2016). In 2016, farmers received up to 222.39 USD/ha to cover crop costs for traditional cover crops, and up to 86.49 USD/ha to grow a rye cover crop (Maryland Agricultural Water Quality Cost-Share Program, 2016).

In addition to federal and state policies and programmes, the potential for private markets or market-driven initiatives can also provide incentives to farmers and ranchers to build soil health and adopt soil health practices. For example in 2009–10, approximately 1,000 ranchers enrolled more than two million hectares of rangeland in voluntary carbon credit trading through the Chicago Climate Exchange (CCX) (Gosnell et al., 2011). The sale of carbon credits from rangeland was developed through the CCX's Rangeland Soil Carbon Offset Protocol. Though the CCX ultimately collapsed, it set a precedent for 'carbon ranching', and a sense of promise that ranchers might be rewarded for the greenhouse gas implications of good stewardship via the private market. Nonetheless, researchers caution that there are a number of challenges associated with such initiatives, including the difficulties of measuring and monitoring soil carbon sequestration, concerns about additionality and permanence in dynamic systems and the fact that net emission reductions are not guaranteed (Booker et al., 2013).

Major buyers of agricultural products, such as grocery stores, restaurant chains and food service companies, can also influence farmer decision-making about soil health practices. The media have focused more heavily on major

Table 4 Examples of soil-health-related Conservation Stewardship Program (CSP) enhancements for United States cropland and rangeland

Cropland	Rangeland
• ENR10: Using Nitrogen (N) provided by legumes, manure or compost • ENR12: Use of legume cover crops as N source • PLT20: High residue cover crops for weed suppression and soil health • SOE05: Intensive no-till • SQL01: Controlled traffic system • SQL04: Use of cover crop mixes • SQL08: Intercropping to improve soil quality and increase biodiversity • SQL12: Intensive cover cropping in annual crops • SQL18: Soil health crop rotation • WQL10: Plant a cover crop that will scavenge residual N • CCR98: Improved Resource Conserving Crop Rotation • CCR99: Resource Conserving Crop Rotation	• ANM03: Incorporate native grasses and/or legumes into 15% or more of herbage dry matter productivity • ANM21: Prairie restoration for grazing and wildlife habitat • ANM37: Prescriptive grazing management • PLT02: Monitoring key grazing areas to improve grazing management • PLT16: Intensive rotational grazing • PLT30: Monitor pasture health using pasture condition scores • SQL16: High species diversity grazing lands • SQL17: Placement of hay feeding areas on low fertility soils • SQL19: Management for rangeland soil health

buyers' influence on changing animal-raising practices, such as the use of gestation grates for sows, the use of cages for egg-laying hens and the use of antibiotics for growth promotion and disease prevention in livestock. Though reducing emissions and fertilizer applications are not inherently synonymous with building soil health, initiatives promoting more sustainable crop production practices which contribute to soil health are also a target of major suppliers and retailers. For example, Walmart is working with suppliers and non-profit organizations such as the Environmental Defense Fund to encourage the adoption of nitrogen management tools and BMPs in crop production, and the Environmental Defense Fund is working with rice, corn and almond growers in California to help them gain access to greenhouse gas emission offset programmes (Pollock, 2015). United Suppliers, a wholesaler and retailer of agricultural inputs in the United States, developed the 'Sustain' platform/programme that requires that farmers adopt some combination of BMPs, such as no-till or cover crops in addition to optimizing and reducing fertilizer application (Betts, 2016). Field to Market, a non-profit organization based in Washington D.C., has also developed a sustainability tool called the Fieldprint® Calculator that helps growers measure and market the sustainability of their operation based upon a number of metrics (Miller, 2014).

7.1 *Designing cost-effective policies for conservation practices in the USA*

When a farmer participates in a federal or state incentive or cost-share programme, economists know that payment to the farmer is at least as large as the farmers' 'willingness to accept' (WTA, or the minimum amount they would need to be paid to be compensated to engage in a practice or change their behaviour). This means that the value of the benefit the farmer receives for implementing the practice less the costs of implementing the practice plus the value of the payment from the programme is greater than zero. However, we do not usually directly observe the costs or benefits of the practice to the farmer, which means that we cannot observe a farmer's WTA for implementing a practice. One potential solution to this issue is to design conservation payment programmes as reverse auctions, where farmers make offers to sell an environmental service or implement a practice or combination of practices in exchange for a value that they bid. Because they are competing against other farmers who may have lower offers, there is an incentive to offer a value close to the farmer's true WTA. However, the way in which reverse auctions are implemented can impact the cost-effectiveness and total environmental benefit of the programme (Claassen et al., 2008; Hellerstein et al., 2015). The U.S. Conservation Reserve Program is an example of a programme that uses a reverse auction design in its general signup, though economists continue to research the economic implications of CRP design and implementation (Hellerstein et al., 2015). This type of research is an example of how behavioural and experimental economics can be used to assess the impact of behavioural nudges on the practice and programme participation, including farmer participation in the conservation practices and programmes.

8 Future trends and conclusion

8.1 *Researching the economics of soil health management systems*

This chapter began by discussing some approaches economists use to empirically model farmer decision-making. In many cases, these approaches focus on modelling the factors that affect a farmer's decision to adopt a single practice - for example, no-till or reduced tillage. By controlling for farmer characteristics, input and output prices, crop choices and other variables, economists take into account the fact that these variables are related in an economic model. However, they are not necessarily and explicitly modelling the fact that the farmer may be making decisions about multiple soil health management practices at once, or how the costs or benefits of implementing one soil health practice may be affected by adopting other practices at the same time. For example, if the farmers are implementing several practices simultaneously as part of a soil conservation plan, the cost and benefit to the farmer of practice implementation depend upon whether the practices being implemented are substitutes or complements. This is also important from a public policy perspective, because it implies that paying farmers to implement some practices may also induce the adoption of other complementary practices, and that for practices that are substitutes, it may be more cost-effective to pay for a practice that is cheaper to implement but provides similar public benefits (Lichtenberg, 2004). Many existing cost-share and incentive programmes that pay farmers to engage in soil health practices pay fixed amounts for the adoption of individual practices, but might be more cost-effective if they considered the existence of complementarity or synergies between practices.

Several previous economic studies look at the farmer's decision to adopt more than one soil-health-related practice (joint or sequential adoption of practices). For example, Khanna (2001) considers the adoption of soil testing and variable rate technology, and finds that farmers with below-average soil quality realize the greatest gains in nitrogen productivity. In addition, her results suggest that for farmers that adopt the technologies, productivity is higher for full-time farm operators. This implies that for these farmers, the adoption of the two technologies might help the farmer to overcome risk aversion and uncertainty about soil quality, thereby reducing over-application of nitrogen. Wu and Babcock (1998) look at farmers' decision-making about tillage, crop rotations and soil testing. The authors find that soil testing alone does not reduce nitrogen application rates, but that a combination of soil testing with crop rotation or conservation tillage does. Future research is needed that looks at the economic and ecological implications of the adoption of multiple practices. For example, how does the cover crop termination method affect the marginal costs of planting a cover crop, and the profitability of no-till production following a cover crop? Does the joint adoption of cover crops and no-till increase herbicide use, and if so, does this have a trade-off in terms of costs of herbicide resistance, or a negative impact on soil biota (and, in turn, soil health)?

8.2 Linking management practices to changes in soil health and economic outcomes

Much previous agronomy and soil science research has focused on linking changes in management practices to changes in yield or profit. In addition, many of these studies are implemented as controlled trials, which makes it difficult to evaluate whether the effects can be generalized to diverse farms and growing conditions. At the same time, economic studies that look at the impact of practice adoption on farms have to control for selection in their studies, or the fact that some farmers are more likely to adopt practices due to variables the researcher cannot observe, such as soil quality or other edaphic variables.

Researchers can develop an understanding of how management practices affect soil health and economic outcomes for farmers by collecting and analysing data over time on management practices, soil health indicators, and economic variables. Such variables include input use and expenditures (e.g. labour, equipment, chemical and water use), crop choice, yields and revenues, and other farm business activities that impact the farm operation's bottom line. It is also important to understand how farmers and land managers make decisions about when to diagnose and treat soil health concerns, and whether soil health monitoring and testing are important tools in this process. Such studies are data-intensive, require data collection over time and necessitate interdisciplinary collaboration. This makes them difficult and costly to conduct. But, without these types of studies being conducted in different locations across the United States and the world, it is difficult to assess the private costs and benefits of implementing soil health practices and how to develop and target public policy incentives that would be appropriate to induce the adoption of practices with public benefits.

8.3 *Valuing the public benefits of changes in soil health at multiple scales*

Linking farm practices and changes in soil health to the provision of public benefits and valuing these benefits require more research. Does a change in a practice on one farm affect the quality of a neighbour's well water, or does it require all farmers within a 10-mile radius to adopt several management practices? If all farmers in the state of Maryland plant a winter cover crop or stop applying manure to fields in the fall, does this have a measurable impact on the water quality of the Chesapeake Bay, and what would be the costs of such a policy? Would the payments needed to induce universal adoption make such a programme cost-effective relative to the expected benefit to water quality?

9 Where to look for further information

U.S. National Agricultural Statistics Service 'Quick Stats' Interface
(www.quickstats.nass.usda.gov)

USDA's National Agricultural Statistics Service provides summary information for the U.S. Agricultural Census and several other surveys via the Quick Stats interface. These include statistics on national- and state-level adoption of several soil health practices, such as cover crops.

Eurostat Agri-Environmental Indicator – Commitments
(http://ec.europa.eu/eurostat/statistics-explained/index.php/Agri-environmental_indicator_-_commitments)
Statistics summarizing coverage of agri-environmental measures in the EU, including land area, expenditures and other measures.

Conservation Technology Information Center
(www.ctic.org)

The Conservation Technology Information Center (CTIC) is a non-profit, public-private partnership that uses public funding to 'champion, promote and provide information on technologies and sustainable agricultural systems that conserve and enhance soil, water, air and wildlife resources and are productive and profitable'. Among other activities, CTIC provides training for industry consultants and professionals, and collects data and conducts surveys about the use of conservation practices, such as no-till agriculture and cover crops.

Sustainable Agriculture Research and Education
(www.sare.org)

Sustainable Agriculture Research and Education (SARE) partners with the University of Maryland and receives funding from the USDA's National Institute for Food and Agriculture to provide information and financial and technical assistance to farmers and ranchers to support sustainable agricultural practices. SARE has published several books on soil health and soil health practices, and has diverse outreach and extension resources available on its website. Since 2012, SARE and CTIC and partners have been conducting an annual survey that asks farmers about their experiences with cover crops.

Center for Behavioural and Experimental Agricultural Research
(www.centerbear.org)
Center for Behavioural and Experimental Agricultural Research is a collaboration among economists at the University of Delaware, Johns Hopkins University and Cornell University to conduct and fund behavioural and experimental research that relates to agricultural and environmental issues. This includes research on USDA conservation programme performance, as well as farmer adoption of conservation and soil health practices.

Soil Health Institute's Research Landscape Tool
(www.soilhealthinstituteresearch.org)

The Soil Health Institute is an independent non-profit organization dedicated to supporting soil health and soil health research. In addition to seeking out and distributing funds for soil health research, the institute has developed a number of resources on its website. One of these resources is a virtual repository of soil health research article abstracts (and links to full publications, though some are only available through subscription or paid services), including research on the economics of soil health.

National Sustainable Agriculture Information Service Case Studies
(www.attra.ncat.org)

The National Sustainable Agriculture Information Service (ATTRA) has published a number of case studies and other resources that address questions about the economics of soil health. In order to find these case studies on their website, go to 'Publications' and click on 'Soils and Compost'.

Landcare Australia
(https://landcareaustralia.org.au/)

Landcare was established as a national programme in Australia in 1989. Its cornerstones are 'being community owned and driven, being bi-partisan in nature and encouraging integrated management of environmental assets, including productive farmland and a sustainable approach to private land management'. The movement has spread to more than 20 other countries globally.

10 References

Al-Kaisi, M., Archontoulis, S. and Kwaw-Mensah, D. 2016. Soybean spatiotemporal yield and economic variability as affected by tillage and crop rotation. *Agronomy Journal* 108: 1267–80.

Arbuckle, J. G. and Ferrell, J. 2012. Attitudes Toward Cover Crops in Iowa: Benefits and Barriers. Extension and Outreach Publications. 22. http://lib.dr.iastate.edu/extension_pubs/22.

Arbuckle, J. G. and Roesch-McNally, G. 2015. Cover crop adoption in Iowa: The role of perceived practice characteristics. *Journal of Soil and Water Conservation* 70(6): 418–29.

Baumgart-Getz, A., Prokopy, L.S. and Floress, K. 2012. Why farmers adopt best management practice in the United States: A meta-analysis of the adoption literature. *Journal of Environmental Management* 96(1): 17–25.

Bennett, L. T., Mele, P. M., Annett, S. and Kasel, S. 2010. Examining links between soil management, soil health, and public benefits in agricultural landscapes: An Australian perspective. *Agriculture, Ecosystems & Environment* 139(1): 1–12.

Betts, L. 2016. Corn and soybean producers find retailer who supports sustainable farming. *Corn and Soybean Digest*. Viewed 5 April 2017. http://www.cornandsoybeandigest.com/corn/corn-and-soybean-producers-find-retailer-who-supports-sustainable-farming

Booker, K., Huntsinger, L., Bartolome, J. W., Sayre, N. F. and Stewart, W. 2013. What can ecological science tell us about opportunities for carbon sequestration on arid rangelands in the United States? *Global Environmental Change* 23(1): 240–51.

Bowman, M., Wallander, S. and Lynch, L. 2016. An economic perspective on soil health. *Amber Waves*. September 2016. https://www.ers.usda.gov/amber-waves/2016/september/an-economic-perspective-on-soil-health/

Carlisle, L. 2016. Factors influencing farmer adoption of soil health practices in the United States: A narrative review. *Agroecology and Sustainable Food Systems* 40(6): 583–613.

Cartwright, L. 2015. Adding cover crops to a corn-soybean rotation. Missouri Cover Crop Economics Case Study 1. Viewed 14 April 2017. https://www.nrcs.usda.gov/wps/PA_NRCSConsumption/download?cid=nrcseprd352827&ext=pdf

Chavas, J. P., Chambers, R. G. and Pope, R. D. 2010. Production economics and farm management: A century of contributions. *American Journal of Agricultural Economics* 92: 356–75.

Claassen, R., Duquette, E. and Horowitz, J. 2013. Additionality in agricultural conservation payment programs. *Journal of Soil and Water Conservation* 68(3):74–8.

Claassen R., Breneman, V., Bucholtz, S., Cattaneo, A., Johansson, R. and Morehart, M. 2004. *Environmental Compliance in Agricultural Policy: Past Performance and Future Potential*. Agricultural Economic Report AER-832, U.S. Department of Agriculture, Economic Research Service, June, Washington, DC. .

Claassen, R., Cattaneo, A. and Johansson, R. 2008. Cost-effective design of agri-environmental payment programs: U.S. experience in theory and practice. *Ecological Economics* 4(1): 737–52.

Claassen, R., Bowman, M., Breneman, V., Wade, T., Williams, R., Fooks, J., Hansen, L., Lovanna, R. and Loesch, C. 2017. Conservation Compliance: How Farmer Incentives are Changing in the Crop Insurance Era. Economic Research Report ERR-234, U.S. Department of Agriculture, Economic Research Service, July.

Conservation Cropping Systems Initiative. 2016. A 321% Return on Investment. *Economics of Conservation Series*. Viewed 14 April 2017. http://ccsin.iaswcd.org/wp-content/uploads/2016/07/CaseStudy-Rulon4.pdf

Cook, R. L. and Trlica, A. 2016. Tillage and fertilizer effects on crop yield and soil properties over 45 years in Southern Illinois. *Agronomy Journal* 108: 415–26.

Cork, S. T., Eadie, L. A., Mele, P. A., Price, R. I. and Yule, D. 2012. The relationships between land management practices and soil condition and the quality of ecosystem services delivered from agricultural land in Australia. Kiri-ganai Research Pty Ltd. Camberra, Australia. 2012 September.

Dale, V. H. and Polasky, S. 2007. Measures of the effects of agricultural practices on ecosystem services. *Ecological Economics* 64(2): 286–96.

Dumanski, J., Reicosky, D. C. and Peiretti, R. A. 2014. Global pioneers in soil conservation: Common elements and lessons learned. *International Soil and Water Conservation Research* 2(1): 1-4.

Dunn, M., Ulrich-Schad, J. D., Prokopy, L. S., Myers, R. L., Watts, C. R. and Scanlon, K. 2016. Perceptions and use of cover crops among early adopters: Findings from a national survey. *Journal of Soil and Water Conservation* 71(1): 29-40.

Foster, A. D. and Rosenzweig, M. R. 1995. Learning by doing and learning from others: Human capital and technical change in agriculture. *Journal of Political Economy* 103(6):1176-209.

Gosnell, H., Robinson-Maness, N. and Charnley, S. 2011. Engaging ranchers in market-based approaches to climate change mitigation: Opportunities, challenges, and policy implications. *Rangelands* 33(5): 20-4.

Hansen, L. and Ribaudo, M. 2008. Economic measures of soil conservation benefits: Regional values for policy assessment. USDA Technical Bulletin No. 1922, 32p, Washington DC .

Hanson, J. C., Lichtenberg, E., Decker, M. and Clark, A. 1993. The profitability of no-tillage corn following a hairy vetch cover crop in the mid-Atlantic region. *Journal of Production Agriculture* 6: 333-437.

Hatfield, J. L., Sauer, T. J. and Prueger, J. H. 2001. Managing soils to achieve greater water use efficiency. *Agronomy Journal* 93(2): 271-80.

Hellerstein, D., Higgins, N. and Roberts, M. 2015. Options for Improving Conservation Programs: Insights from Auction Theory and Economic Experiments. Economic Research Report No. (ERR-181), 53p, Washington DC.

Hudson, B. 1994. Soil water and available water capacity. *Journal of Soil and Water Conservation* 49(2): 189-94.

Huguenin, M. T., Leggett, C. G. and Paterson, R. W. 2006. Economic valuation of soil fauna. *European Journal of Soil Biology* 42: S16-S22.

Karlen, D. L., Cambardella, C. A., Kovar, J. L. and Colvin, T. S. 2013a. Soil quality response to long-term tillage and crop rotation practices. *Soil and Tillage Research* 133: 54-64.

Karlen, D. L., Kovar, J. L., Cambardella, C. A. and Colvin, T. S. 2013b. Thirty-year tillage effects on crop yield and soil fertility indicators. *Soil and Tillage Research* 130: 24-41.

Kemper, W. D., Schneider, N. N. and Sinclair, T. R. 2011. No-till can increase earthworm populations and rooting depths. *Journal of Soil and Water Conservation* 66: 13A-17A.

Khanna, M. 2001. Sequential adoption of site-specific technologies and its implications for nitrogen productivity: A double selectivity model. *American Journal of Agricultural Economics* 83(1): 35-51.

Knight, L. 2016. Developing Conservation Case Studies for Decision-making. Economics Technical Note 200-ECN-4. USDA Natural Resources Conservation Service. Viewed 10 July 2017. https://www.nrcs.usda.gov/wps/PA_NRCSConsumption/download?cid=nrcseprd1298424&ext=docx

Lal, R. 2004. Soil carbon sequestration impacts on global climate change and food security. *Science* 304(5677): 1623-7.

Lichtenberg, E. 2004. Cost-responsiveness of conservation practice adoption: A revealed preference approach. *Journal of Agricultural and Resource Economics* 29(3): 420-35.

Lichtenberg, E., Hanson, J. C., Decker, M. and Clark, A. 1994. The profitability of legume cover crops in the mid-Atlantic region. *Journal of Soil and Water Conservation* 49: 582-5.

Maryland Agricultural Water Quality Cost-Share Program, 2016. Opening doors for Maryland's Conservation Farmers: MACS Annual Report. http://mda.maryland.gov/resource_conservation/counties/MACS_FINAL.pdf.

Mausbach, M. J. and Dedrick, A. R. 2004. The length we go. *Journal of Soil and Water Conservation* 59(5): 96–103.

Meitner, K. 1999. Making the most of it! *No-Till Farmer*. Viewed 11 July 2017. https://www.no-tillfarmer.com/articles/3739-making-the-most-of-it?v=preview

Miller, S. 2014. Fieldprint calculator uses a USDA tool to help farmers track water quality improvement. *USDA Blog*. Viewed 6 April 2017. https://www.usda.gov/media/blog/2014/04/16/fieldprint-calculator-uses-usda-tool-help-farmers-track-water-quality

No-Till Farmers. 2017. National No-Tillage Conference. Viewed 11 July 2017. https://www.no-tillfarmer.com/nntc

Pollock, D. 2015. Pilot project to generate greenhouse gas credits from almonds, corn. *Western Farm Press*. Viewed 8 April 2017. http://www.westernfarmpress.com/tree-nuts/pilot-project-generate-greenhouse-gas-credits-almonds-corn

Prokopy, L. S., Floress, K., Klotthor-Weinkauf, D. and Baumgart-Getz, A. 2008. Determinants of agricultural best management practice adoption: Evidence from the literature. *Journal of Soil and Water Conservation* 63(5): 300–11.

Ransom, M. 2014. Case Study: Transition to Cover Crop for a Beginning Farmer. Viewed 10 April 2017. https://www.nrcs.usda.gov/wps/PA_NRCSConsumption/download?cid=stelprdb1256123&ext=pdf

Rawls, W. J., Pachepsky, Y. A., Ritchie, J. C., Sobecki, T. M. and Bloodworth, H. 2003. Effect of soil organic carbon on soil water retention. *Geoderma* 116(1): 61–76.

Reimer, A. P., Weinkauf, D. K. and Prokopy, L. S. 2012. The influence of perceptions of practice characteristics: An examination of agricultural best management practice adoption in two Indiana watersheds. *Journal of Rural Studies* 28(1): 118–28.

Ribaudo, M., Greene, C., Hansen, L. and Hellerstein, D. 2010. Ecosystem services from agriculture: steps for expanding markets. *Ecological Economics* 69(11): 2085–92.

Robinson, D. A., Hockley, N., Dominati, E., Lebron, I., Scow, K. M., Reynolds, B., Emmett, B. A., Keith, A. M., de Jonge, L. W., Schjonning, P., Moldrup, P., Jones, S. B. and Tuller, M. 2012. Natural capital, ecosystem services, and soil change: why soil science must embrace an ecosystems approach. *Vadose Zone Journal* 11(1).

Rogers, E. M. 2003. *Diffusion of Innovations*, 5th Edition. Free Press, New York, NY.

Schoengold, Karina, Ding, Y. and Headlee, R. 2015. The impact of AD HOC disaster and crop insurance programs on the use of risk-reducing conservation tillage practices. *American Journal of Agricultural Economics* 97(3): 897–919.

Snapp, S. S., Swinton, S. M., Labarta, R., Mutch, D., Black, J. R., Leep, R., Nyiraneza, J. and O'Neil, K. 2005. Evaluating cover crops for benefits, costs and performance within cropping system niches. *Agronomy Journal* 97(1): 322–32.

Soil and Water Conservation Society. 2017. 2017 National Conference on Cover Crops. Viewed 10 July 2017. http://www.swcs.org/en/conferences/2017_national_conference_on_cover_crops/

Sunding, D. and Zilberman, D. 2001. Chapter 4: The agricultural innovation process: Research and technology adoption in a changing agricultural sector. . In: *Handbook of Agricultural Economics*, Volume 1 (Part A), North Holland, pp. 207–61.

Swinton, S., Frank Lupi, Philip Robertson, G., Hamilton, S. K. 2007. Ecosystem services and agriculture: Cultivating agricultural ecosystems for diverse benefits. *Ecological Economics* 64(2): 245–52

Tallman, S. 2012a. No-till case study, Brown's Ranch: Improving soil health improves the bottom line. National Sustainable Agriculture Information Service, National Center for Appropriate Technology. Viewed 6 April 2017. https://attra.ncat.org/attra-pub/download.php?id=414

Tallman, S. 2012b. No-till case study, Bauer Farm: Cover crop cocktails on former CRP land. National Sustainable Agriculture Information Service, National Center for Appropriate Technology. Viewed 6 April 2017. https://attra.ncat.org/attra-pub/download.php?id=420

Tallman, S. 2012c. No-Till Case Study, Richter Farm: Cover Crop Cocktails in a Forage-Based System . National Sustainable Agriculture Information Service, National Center for Appropriate Technology. Viewed 14 July 2017. https://attra.ncat.org/attra-pub/download.php?id=417

Tomer, M. D. and Locke, M. A. 2011. The challenge of documenting water quality benefits of conservation practices: A review of USDA-ARS's conservation effects assessment project watershed studies. *Water Science and Technology* 64(1): 300–10.

USDA Natural Resources Conservation Service. 2016. Profiles in Soil Health. Viewed 14 April 2017. https://www.nrcs.usda.gov/wps/portal/nrcs/detail/national/soils/health/?cid=stelprdb1083183

USDA Natural Resources Conservation Service, New York. 2016. Economic Case Study – Farmer Profile, Dave Magos. Viewed 3 April 2017. https://www.nrcs.usda.gov/wps/portal/nrcs/detail/ny/soils/health/?cid=nrcseprd1298872

USDA Natural Resources Conservation Service. 2017. Healthy, productive soils checklist for growers. Fact Sheet. Viewed 10 July 2017. https://www.nrcs.usda.gov/wps/PA_NRCSConsumption/download?cid=nrcseprd1318014&ext=pdf

U.S. Department of Agriculture, Natural Resources Conservation Service. 2016. Effects of Conservation Practice Adoption on Cultivated Cropland Acres in Western Lake Erie Basin, 2003–06 and 2012, 120p https://www.nrcs.usda.gov/Internet/FSE_DOCUMENTS/nrcseprd889806.pdf.

Wade, T., Claassen, R. and Wallander, S. 2015. Conservation-Practice Adoption Rates Vary Widely by Crop and Region. Economic Research Service EIB 147, Washington DC., 40p.

Wallander, S., Aillery, M., Hellerstein, D. and Hand, M. 2013. The Role of Conservation Programs in Drought Risk Adaptation. Economic Research Service ERR 148, Washington DC.

Watkins, B., Anders, M. M. and Windham, T. E. 2004. An Economic Comparison of Alternative Rice Production Systems in Arkansas. *Journal of Sustainable Agriculture* 24(4): 57-78.

Wu, J. J. and Babcock, B. A. 1998. The choice of tillage, rotation, and soil testing practices: Economic and environmental implications. *American Journal of Agricultural Economics* 80 (3): 494–511.

Zimmermann, A. and Britz, W. 2016. European farms' participation in agri-environmental measures. *Land Use Policy* 50: 214–28.

Chapter 6

The economics of precision agriculture

James Lowenberg-DeBoer, Harper Adams University, UK

1 Introduction

While precision agriculture (PA) has enormous potential for reducing the environmental footprint of farming, awareness is growing with respect to the social implications of the technology and the widespread adoption of PA depends on economics. Farming is a business and technology is adopted if it provides benefits for the farmer and farm household. Sometimes those benefits are qualitative (e.g. more time for recreation, reduced fatigue, less stress), but often those benefits are in monetary terms (i.e. more stable cash flow, higher and less variable profits). Farming is subject to what has been called 'economic physics' in the sense that someone must pay the bills regardless of the ideological framework within which the farmer is functioning just as everyone is subject to gravity whether or not they approve or think it is fair. Labour and capital must be compensated. The opportunity costs of land and other natural resources must be covered. That compensation might be in cash or in kind. Consequently, the objective of this chapter is to summarize the state of the art in economics of PA. Assuming that adoption is one of the best indicators of technology profitability and benefits, the topic will be approached by (1) reviewing the status of PA adoption worldwide, (2) assessing the degree to which PA adoption was predicted by economic studies and (3) describing future adoption as predicted by recent PA economics research. The chapter will be of interest to PA researchers, research administrators seeking to identify fruitful research areas, educators, farmers and agribusinesses.

http://dx.doi.org/10.19103/AS.2017.0032.19

Many definitions have been provided for 'precision agriculture' (e.g. Robert et al., 1994; National Research Council, 1997; Lowenberg-DeBoer and Swinton, 1997; Whelan and McBratney, 2000; Thorp, 2014), but most centre on use of electronic information technology to better manage spatial and temporal variability in crops, livestock, forestry and other biological systems. PA is not one technology, but a toolbox of technologies. The 'classic' PA technologies originally proposed in the early 1990s were yield monitoring, remote sensing with satellites or aircraft, and variable rate technology (VRT) for fertilizer, pesticide and seed, supported by use of the Global Positioning System (GPS) for location finding and Geographic Information Systems (GIS). In the late 1990s, Global Navigation Satellite System (GNSS) guidance was introduced first in the form of lightbars and later as autosteer. GNSS guidance was soon joined by related technologies like sprayer boom control and planter row shutoffs. At the time of writing, improved GNSS (the United States, European, Russian and Chinese systems), artificial intelligence (AI), analysis of 'big data' and robotics are being added to the PA toolbox.

This analysis will focus on PA for crops because that has received the most attention by economists, but the principles outlined can be applied to livestock, forestry and other biological production systems. The analysis will concentrate on the use of PA for mechanized agriculture because that is where most PA technology has been deployed thus far, but observations will be made about adapting PA technology for small- and medium-sized farms in developing countries. PA has implications for the whole food value chain, from input supply to processing and marketing, but this chapter will concentrate on the primary production stages. The main focus of this chapter will be on 'action' technologies like GNSS guidance and VRT that make a physical change in crop management (e.g. GNSS guidance reduces pass to pass overlap, VRT applied inputs where and when they are needed). Many of the technologies in the PA toolkit are information-gathering tools (e.g. yield monitoring, satellite images, drones). The profitability of information gathering depends on the economics of the action technologies that implement decisions made with the information gathered.

The beginnings of PA are often dated to the release in the United States of GPS for civilian use in the late 1980s. The first published PA economic studies are from the mid-1990s. In the early days of PA, at least two quasi-exhaustive reviews of the economics research were conducted. Lambert and Lowenberg-DeBoer completed a review in 2000. Griffin et al. (2004) updated that review. Those two reviews found that about two-thirds of research articles reported benefits from PA technology, but only 20% of the articles included management, learning and human capital costs in their profitability estimates. The two reviews identified better returns to PA for crops that are more intensively managed and have a higher gross revenue per unit area. For example, the reviews show that returns to PA for cotton, sugar beets and corn were relatively high, but low for extensive wheat production in the US Great Plains. The reviews also showed mixed results for VRT fertilizer. At the time of the 2000 and 2004 PA economics reviews, GNSS guidance for agricultural equipment had only recently been commercialized in the United States. The few studies of GNSS guidance profitability had identified it as a highly profitable PA technology. Both reviews also highlighted the variability of profit estimates depending on the combination of PA technologies and the economic methodology used.

After 2004, the number of economic studies exploded and overall reviews of PA economics became more difficult. Partial reviews of the recent PA economics literature can be found in Pierpaoli et al. (2013), Griffin et al. (2017a) and Balafoutis et al. (2017). There were some reviews of specific PA technologies and technology packages. For example, Tenkorang and Lowenberg-DeBoer (2008) reviewed hundreds of agricultural remote sensing articles and reports. They found only a few that reported economic benefits and an even smaller number that explained how that economic benefit was estimated. For twelve studies they found enough information to apply standard budgeting methods and found that remote sensing increased profit estimates by about US$31.74/ha. They argued for more detailed reporting of budget assumptions and replicating studies over several years and at multiple geographic locations.

As governments became aware of precision agriculture, there were numerous policy-oriented studies which attempted to identify the economic benefits of PA. For example, in 2014 the EU Directorate for Internal Policies reviewed a scattering of studies from Europe and elsewhere in the world (Zarco-Tejada et al., 2014). They found that there seemed to be modest, but consistent benefit, to GNSS guidance almost everywhere and that controlled traffic using GNSS guidance may be the single biggest PA benefit for EU agriculture, but the profitability of VRT depended on the crop, soil conditions and many other factors. The OECD (2016) and the UK Agriculture and Horticulture Development Board (Awan, 2016) reported similar results. The US Department of Agriculture (USDA) reported that PA benefits for corn producers were positive, but small (Schimmelpfennig, 2016). This was based on an in-depth analysis of data from the USDA Agricultural Resource Management Survey (ARMS) data.

In spite of the massive literature on the economics of precision agriculture, questions about the profitability of PA continue. Complaints about the slow adoption are a regular part of almost every analysis of the potential for PA and that slow adoption is often linked to uncertainty about benefits and profitability. For example, Lowenberg-DeBoer (1998) focused on the 'uneven' adoption path of some PA technologies because they were coming on the market in an immature form and before supporting services were available. Cook et al. (2000) noted that slow adoption of PA by Australian farmers was in part due to perceived lack of benefit. Zhang et al. (2002) pointed out that in spite of substantial research investment in PA in the United States, Canada, Australia and the European Union, only a small portion of farmers in those countries use the technology. Kitchen (2008) noted that precision agriculture adoption has been hindered because the agronomy and engineering have not come together in ways that compensate the farm investment of time and money. Tey and Brindal (2013) review some 50 papers focused on the adoption of PA technologies and find that perceived profitability of PA significantly influences adoption decisions. The economic analysis of PA adoption in Nebraska by Castle et al. (2017) was motivated by lack of clarity related to the profitability of the technology.

Given the continuing questions about the profitability and other economic benefits of PA, the overall goal of this chapter and the extant literature on the economics of PA, the specific objectives of this chapter are to:

1 review the adoption of PA technology worldwide,
2 identify the degree to which economics research has correctly identified technologies that will be widely adopted and
3 overview the current (last decade) PA economics research and what it suggests as the likely PA adoption trends of the next decade.

2 Adoption of PA technology

Adoption of PA technology must be inferred from sparse data. PA adoption is not tracked regularly by any national agricultural statistics service. For example the USDA collects information on technology use as part of the ARMS which focuses on one crop at a time, with irregular intervals between surveys. Consequently since 2000, there is PA technology use data for corn in 2000, 2005 and 2010; rice: 2006 and 2013; cotton: 2000, 2003 and 2007; peanuts: 2004 and 2013; soybeans: 2000, 2002, 2006 and 2012; and spring wheat: 2000, 2004 and 2009. Some countries do occasional surveys, like the DEFRA farming practices survey in the United Kingdom in 2012. In many countries, PA adoption data depends on one-of-a-kind surveys by researchers or information from manufacturers and agribusiness. Understanding trends in the PA adoption requires piecing together information from disparate sources and irregular intervals.

The best historical time series of PA adoption data is from the CropLife/Purdue dealership survey which has been done regularly since the mid-1990s (Erickson et al., 2018). This survey has been supported by Meister Publication/CropLife magazine and depends on their almost exhaustive list of US agricultural input suppliers. There have typically been 200–300 responses, mostly from agricultural input supply firms. Responses have been concentrated in the Midwest. From 1996 to 2015, it was purely a paper survey sent through the post. In 2017, a paper copy was still mailed, but a web-based survey was also made available to respondents. From 1996 to 2009, the survey was done annually. After 2009, it was done biennially. The survey is not a panel; the set of respondents differs from year to year so some sampling variability is evident. The survey has typically asked only if a technology was used, not the intensity of use. If two dealership both have ten fertilizer applicators, one dealership has GNSS autosteer on all ten and the other has autosteer only on one of the ten, both are still counted as using GNSS autosteer. Clearly, the CropLife/Purdue survey is not a perfect source of PA adoption data, but it is relatively consistent over time with some of the same question asked over many years. A survey of agricultural input suppliers is particularly relevant for PA adoption in the United States, because many American farmers access PA technology through hiring services rather than by owning and operating the equipment directly.

For example, Fig. 1 shows the use of GPS guidance and offering of sensor services by US agricultural input dealerships. The CropLife/Purdue data shows the adoption of GNSS guidance clearly. GNSS light bars were introduced for ground-based agricultural equipment in the United States in the late 1990s and by 2004 over 60% of the dealerships were using them. By 2009, almost 80% of dealerships were using light bars. GNSS autosteer was introduced into the US Midwest about 2003. The CropLife/Purdue survey tracks the growth in autosteer use from only 5% in 2004 to around 80% in 2015 and 2017. The decline in light bar use after 2009 seems to show a case of technology replacement, in which dealerships gained experience with guidance through lightbars, but after 2009 have replaced lightbars with autosteer. The decline in autosteer use from 83% in 2015 to 78% in 2017 is probably more due to differences in the sample, than it is to any dis-adoption of this technology. The growth in sprayer section control is also documented in the rise from 39% in 2011 to the mid-70% in 2015 and 2017. Clearly, GNSS guidance is a widely adopted PA technology and it was

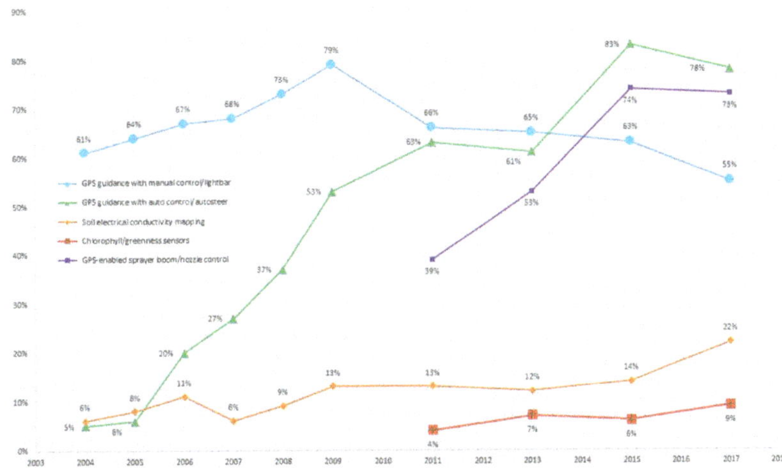

Figure 1 GPS guidance and sensor services at US agricultural dealerships, 2004–17. Source: Erickson et al. (2018).

adopted relatively quickly. In the case of both lightbars and autosteer, within about 10 years of introduction, GNSS was used by around 80% of the dealers. The CropLife/Purdue data indicate that complaints about slow adoption of PA technology should not apply to GNSS guidance.

The sensor data in Fig. 1 show different adoption patterns. The number of dealerships offering electrical conductivity mapping rose very gradually from 6% in 2004 to 22% in 2017. Chlorophyll/greenness sensors used in nitrogen management rose from 4% in 2011 to 9% in 2017. The CropLife/Purdue data indicate that the slow adoption complaint could apply to PA sensors.

Figure 2 shows the VRT services offered by the US agricultural input supply dealerships. Farmer use of these services in the United States is described below. A few agricultural input dealers started offering VRT fertilizer in the early 1990s. By 1999, the percentage offering that service was over 30%. That percentage has risen gradually to almost 80%. In the United States agricultural lime application is often done by a different set of contractors with specialized equipment that can spread crushed limestone and so it is tracked separately from granulated dry fertilizer in this survey. Controller VRT lime has followed the same pattern as VRT fertilizer, but 10–15% lower. The use of VRT lime is probably underestimated by the CropLife Purdue survey because it does not separately track application of pelleted lime or 'manual' site-specific lime application. Pelleted lime can be applied by VRT fertilizer equipment along with other granulated products. Also in many US soils, pH differences often are observed in relatively large areas corresponding to soil type or topographical

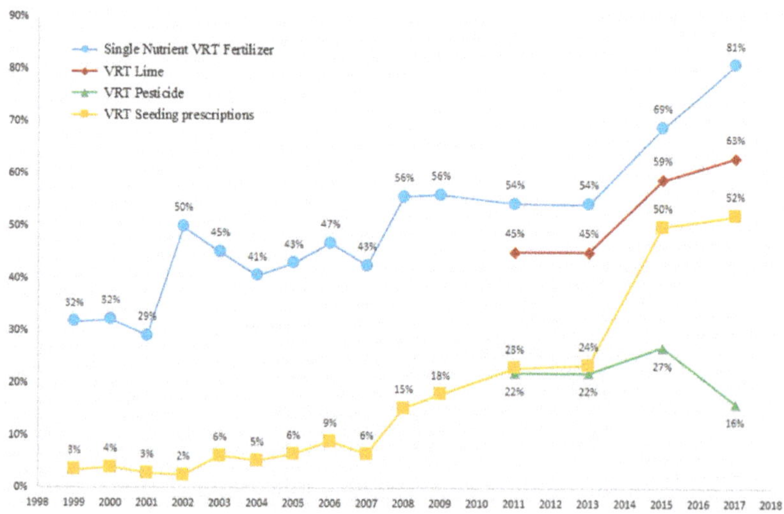

Figure 2 VRT services at US agricultural dealerships, 1999–2017. Source: Erickson et al. (2018).

position. On those soils, farmers sometimes manually mark the area to be limed (e.g. by disking or ploughing around the designated area).

In the United States agricultural input suppliers often sell seed, but they rarely provide seeding services. Most farmers have their own planters and many of those planters are capable of varying seeding rates by management zone. The VRT seeding line in Fig. 2 shows the growing interest in 'prescription seeding'. In this case, the agricultural input dealer works with the farm to develop a map specifying seeding rates and sometimes varieties by management zone.

In spite of the irregular survey schedule and the crop focus, rather than a farm focus, the USDA ARMS data is probably the best information available anywhere in the world on farmer adoption of PA technology (USDA, 2018). Compared to the CropLife/Purdue results, the USDA data tells the same story about GNSS guidance and a different story about VRT. Figure 3 shows a scatter plot of the USDA ARMS estimates of GNSS guidance use by farmers producing six different crops: corn, soybeans, winter wheat, rice, cotton and peanuts. The USDA data does not distinguish between lightbar manual guidance and autosteer. The adoption estimates for 2001 and 2002 predate general commercial availability of autosteer in the United States and are mostly manual lightbar use. The more recent estimates are a mix of lightbar and autosteer adoption. The USDA guidance adoption estimates show a very clear trend that is very similar to the dealership adoption pattern in Fig. 1, especially the GNSS autosteer line. Dealerships probably started adopting lightbars earlier than

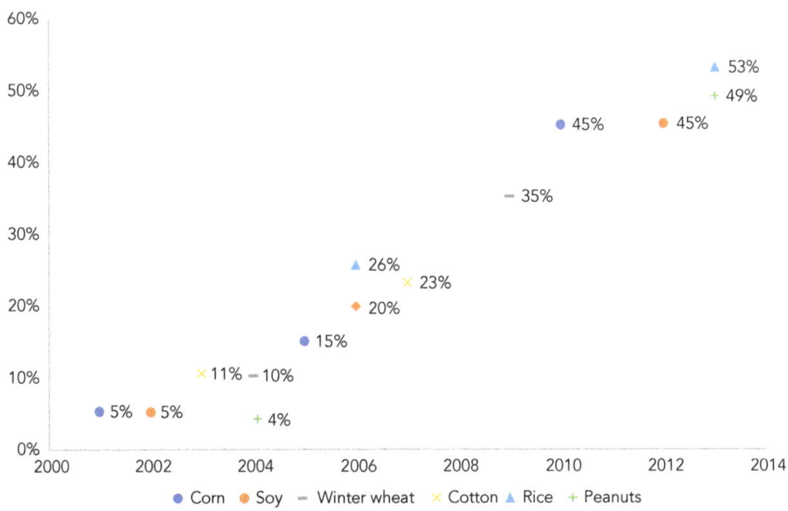

Figure 3 Percentage of planted area by crop on which GPS guidance or autosteer was used, 2000-13. Source: Based on data from USDA ARMS - https://data.ers.usda.gov/reports.aspx?ID=17883

farmers and so by 2004, when the question was added to the CropLife/Purdue survey, the adoption rate was already over 60%.

Figure 4 is a scatter plot of VRT use for any purpose by farmers producing six crops. It shows a broad scatter of points mostly below 20% of planted area with only slight evidence of a trend towards greater adoption. In those years when data on the purpose of VRT is available, it shows that it is mostly for fertilizer application. This slight trend towards greater adoption of VRT is in spite of heavy media coverage of VRT technology, strong farmer interest in the technology and government subsidies for VRT in some areas. The slight upward trend in VRT occurred in the 2008–13 period of relatively high farm crop prices. Some evidence suggests that when crop prices declined sharply after 2014, farmers cut back on VRT and the intensive soil sampling on which it is usually based (Erickson et al., 2018).

The sparse data from many different parts of the world confirm the rapid and widespread adoption of GNSS guidance for mechanized agriculture and the relatively slow and irregular pace of VRT adoption. For example:

- In a one-time survey of agricultural input supply dealerships in Ontario, Canada, Mitchell et al. (2017) found that 74% of the dealerships used autosteer, 66% used section boom control and 88% offered VRT services, but VRT was used on only 28% of the crop area where they provided custom fertilizer spreading. The same survey found that lightbars were used by 30% of respondents.

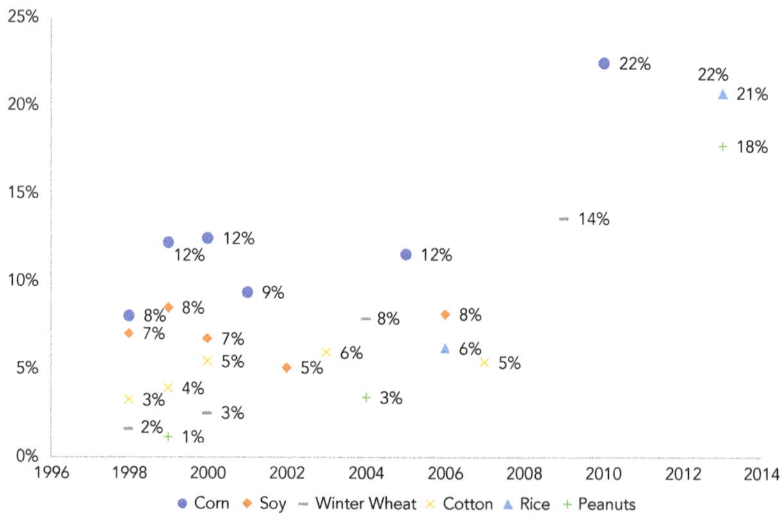

Figure 4 VRT for any purpose 1998 to 2013, Corn, Soybean, Winter Wheat, Cotton, Rice and Peanuts. Source: Based on data from USDA ARMS - https://data.ers.usda.gov/reports.aspx?ID=17883

- Miller et al. (2017) showed a similar pattern of adoption among farmers in the US state of Kansas using data from the Kansas Farm Management Association (KFMA). Use of GPS autosteer grew quickly from commercial introduction in the early 2000s to 66% by 2015, while VRT fertilizer grew slowly from the mid-1990s to about 25% in 2015. In this dataset, use of GNSS lightbars plateaus at about 40% in 2010. It should be noted that farm management association members are usually more interested in technology and improving farm management than the average US farmer and, consequently, may show somewhat higher adoption levels in the early stages of the adoption process.
- In the US state of Nebraska, Castle et al. (2017) studied technology adoption among farmer clients of Nebraska Farm Business, Inc. (NFBI). Over about a decade, use of GNSS autosteer among this group grew to over 70%, while use of VRT N, P, K and lime grew to about 50% in the two decades from the mid-1990s to 2014. Lightbar use peaked in this data at about 35% in 2010 and has declined slightly since then. The 50% VRT fertilizer adoption is high by worldwide standards and may be related to willingness to experiment among NFBI clients. Like the KFMA, NFBI members tend to be more interested in technology and improved management than the average US farmer.
- In Brazil, a phone survey of 922 large-scale grain producers showed that the majority used GNSS guidance and many used VRT fertilizer based on relatively large grid sizes (Molin, 2016). About 23% of them reported using autosteer and 14% lightbars. Thirteen percent reported using a VRT controller for fertilizer or pesticide.
- The National Agricultural Research Institute (INTA) of Argentina reports PA adoption in terms of number of machines instead of percent of farmers or percent of cropland, but the pattern is similar to that observed in the United States and Canada (Méndez et al., 2014; Scaramuzza et al., 2017). The number of GNSS guidance units is much greater than the number of VRT machines. The number of GNSS lightbars used for sprayers in Argentina has increased from 10 in 1998 to 18 342 in 2015. The number of GNSS autosteer units increased from 3 in 2004 to 5767 in 2015. The number of self-propelled variable rate urea and ammonium nitrate solution applicators increased from 50 in 2005 to only 854 in 2013. The number of seeders capable of variable rate starter fertilizer rose from 3 in 1997 to 3263 in 2015.
- In Australia, Llewellyn and Ousman (2014) found that 77% of Australian grain growers were using autosteer. The same survey found that while almost half of the Australian growers were varying fertilizer rates within fields, only about 15% were using computer controlled variable rate technology. This survey was done in 2012 in the main grain growing regions of Australia.

- In the United Kingdom in 2012 a nationwide survey of farming practices found that 22% of farmers were using GNSS guidance and 16% were using VRT (DEFRA, 2013), with higher adoption of both GNSS guidance and VRT on larger and more grain-focused farms. An internet survey in 2016 found similar adoption levels though the sample was dominated by younger respondents (Pickthall and Trivett, 2017).
- Elsewhere in Europe the PA adoption data is not recent, but shows the same pattern. A farmer survey documented that in Germany in 2010, 32% of respondents used GNSS guidance and 8% used VRT (Lawson et al., 2011). The same survey showed that in Denmark at that time 9% of respondents used GNSS guidance and 4% use VRT. In Finland, 4% use GNSS guidance and 3% used VRT.

While GNSS guidance is being adopted quickly almost wherever agriculture is mechanized, VRT is more likely to be found in 'hot spots' where the profit potential and soil variability combine to motivate adoption. The classic example of a VRT adoption hot spot is sugar beet producers in the Red River Valley of the North in the US states of Minnesota and North Dakota. Franzen (2017) documents that 53% of the sugar beet area in the Valley was managed with VRT nitrogen and the grid soil sampled area was even larger. Because soils in the Valley are originally an ancient lakebed, they are relatively uniform and some grid sampled fields were not judged variable enough to justify VRT application. Llewellyn and Ousman (2014) note that while overall adoption of computer controlled VRT in Australia is only 15% of grain farmers, it is 37% in the Victoria Mallee area where within-field variability is particularly noticeable.

Like GNSS guidance, adoption of other PA technology that depends only on location information is quite high. For example, the 2017 CropLife/Purdue survey found that 73% of responding dealerships used GNSS sprayer boom control. In the 2017 Ontario dealership survey, 69% use GNSS sprayer boom control and 25% use sprayer turn compensation. In Argentina in 2016, there were 9035 autosteer units and 3375 sprayers equipped with boom control. Since sprayer boom control, turn compensation and seed row shutoffs were introduced commercially quite recently, the USDA ARMS data and most other PA surveys do not yet include questions regarding these technologies.

PA information-gathering technology is often used to guide VRT and like VRT it has a mixed adoption picture.

- With VRT fertilizer, combine harvester yield monitors were part of the original PA technology introduced in the early 1990s. In industrialized countries, yield monitors have become standard equipment on many new combines, but many of those yield monitors are not used at all or used without GNSS so that the data cannot be mapped. For example the

USDA ARMS data shows that for corn in 2010, 61% of respondents used a combine equipped with a yield monitor, but only 34% made yield maps. In 2006, 45% of US soybean farmer respondents used a combine equipped with a yield monitor, but 20% made yield maps. For rice in 2006, 29% of respondents used a yield monitor, but only 10% made maps. For winter wheat in 2009, 36% of respondents used a yield monitor, but only 6% made maps. Farmers worldwide complain that they often do not know how to use the yield map data and consequently lack the motivation to collect good quality yield data for maps.

- Efforts to use satellite remote sensing data for agricultural management has been made since the 1970s and some agricultural service providers have adopted the technology, but regular use of that data by farmers is still not widespread. The 2017 CropLife/Purdue survey showed that 59% of dealerships offered satellite or aerial imagery as a service to farmers. The 2017 Ontario survey revealed that 42% of the dealerships used satellite or aerial imagery for internal dealership decision-making. The most recent USDA ARMS data on the subject shows that in 2005 only 6% of corn farmers used an aerial or satellite image, in 2002 only 1% of soybean farmers, in 2003 7% of cotton farmers and in 2004, 3% of wheat farmers. Lawson et al. (2011) indicated that aerial images were used in 2010 by 10% of the German respondents, 1% of the Danish respondents and none of the Finnish respondents.
- Interest in drones and unmanned aerial vehicles has grown substantially. PA conferences and scientific journals are packed with drone presentations and articles. Some dealerships have started offering drone imaging services. The 2017 CropLife/Purdue survey showed that 32% offer drone image services. The 2017 Ontario dealership survey indicated that 21% offered drone imagery for fertilizer prescription map development.
- In the United States and Canada most agricultural input suppliers offer some kind of intensive soil sampling, usually either grid or zone based. The 2017 CropLife/Purdue survey showed 82% of dealerships offering this service. The 2017 Ontario survey showed that 81% of the dealerships offered this service. The USDA ARMS data suggests that farmer's use of intensive soil sampling data for maps is substantially less than the service offerings: corn farmers, 2010, 22%; soybean producers, 2012, 19%; rice farmers, 2013, 12%; cotton farmers, 2007, 5%; peanut growers, 2013, 24%; and winter wheat producers, 2009, 6%. In the United Kingdom, the 2012 DEFRA survey showed 14% of farmers using some type of precision agriculture-based soil map. Lawson et al. (2011) indicated that German respondents reported 14% using grid sampling and 5% zone sampling, Danish respondents reported 5% using grid sampling and 1% zone sampling and Finnish respondent reported no grid sampling and 1%

using zone sampling. Molin (2016) reports that 79% of the Brazilian large-scale farmers reported doing PA soil mapping, but the coarse scale of the sampling (often over 5 ha) raises questions about the accuracy of those soil maps.

- Soil electrical conductivity (EC) measurement for nitrogen management is a technique that predates the PA era, but dealership EC services offerings are modest and overall farmers have not been enthusiastic users of those services. The 2017 CropLife/Purdue survey found that 34% of US dealerships offered EC soil mapping. The 2017 Ontario survey found that 11% of the dealerships offered EC mapping. USDA ARMS surveys from 2002 for soybean, 2003 for cotton and 2010 for corn showed about 1% of respondents using EC-based soil maps. Farmers and agronomists complain about difficulty of interpretation of soil EC.
- Griffin et al. (2017b) used data from the KFMA to show that VRT fertilizer application is more likely to be adopted in combination with yield monitoring and intensive soil sampling.

Overall, adoption by farmers and agribusiness of GNSS guidance and other technologies that depend primarily on location information is strong worldwide. The story on adoption of technologies that depend on a broader array of information is mixed. VRT fertilizer technologies have been adopted in some niches where the economic motivation is compelling and the practical constraints have not been overwhelming. The use of VRT N by sugar beet producers in the Red River Valley of Minnesota and North Dakota, USA, is a good example of strong VRT niche adoption. More generally for grain and oilseed producers, it is difficult to find adoption of information-intensive VRT fertilizer technology that is more than 20% of farmers or of cropland. Adoption of VRT for pesticides is even more limited. VRT seeding is growing fast, but is not yet widespread. The next section considers the degree to which economic studies have predicted the observed PA adoption pattern.

3 PA adoption and economics

In general, the current PA pattern was accurately predicted by economic studies done in the 1990s and in the early years of the twenty-first century. Those studies that looked beyond partial budgeting of variable cost and revenue changes, to include changes in management costs, training expense and personnel-related issues were particularly useful. Since GNSS guidance was introduced for ground-based agricultural equipment in the late 1990s, almost all economic studies have shown positive economic benefits which could be quantified and substantial qualitative benefits which were more difficult to measure. Consequently, the rapid adoption of this and related technologies should not come as a surprise. Since the early 1990s, VRT fertilizer economics studies have shown mixed results and the adoption record reflects that mixed picture with strong adoption in some niches, but with widespread adoption awaiting technical breakthroughs that reduce costs and enhance effectiveness. This section will review the track record of those early PA economic studies.

The first published study of GNSS guidance economics was done at Purdue University with data supplied by the companies introducing the initial products (Lowenberg-DeBoer, 1999). The study focused on the benefits of GNSS lightbar systems because autosteer was not yet on the North American market. It summarized the qualitative benefits of GNSS lightbar guidance and quantified the benefits of reduced skip and overlap minus operating costs compared to the foam and disk markers used up to that time. It showed that for custom operators, reduction in skip and overlap would pay for GNSS guidance and show a modest benefit. The situation was similar for the many farmers who had already invested in GNSS for yield monitoring. The study predicted the growth of GNSS autosteer technology in agriculture based on technology that was already being used in construction and mining at that time.

Subsequent studies added detail to the benefit estimates and the potential of related technologies. For example, Watson and Lowenberg-DeBoer (2004) added autosteer technology to the economic assessment, and considered the potential for farm expansion and controlled traffic. They concluded that autosteer would be profitable for a substantial group of US farmers. Griffin et al. (2005) put the GNSS guidance decision in a whole farm context and evaluated timeliness benefits for guidance. Batte and Ehsani (2006) added sprayer boom section control to the analysis and considered the impact of irregular field shape and field obstructions like waterways, trees and rocks. Bergtold et al. (2009) estimated benefits of autosteer in cotton production. Ortiz et al. (2013) estimated benefit of autosteer for peanut production.

The mixed results of VRT fertilizer economics studies are in contrast to the consistently positive results found in GNSS guidance studies. One of the earliest reviews of VRT fertilizer economics was done by Swinton and Lowenberg-DeBoer (1998). They reviewed nine VRT fertilizer studies with economic analysis done in the 1987–96 period. Because budgeting assumptions in the studies varied widely across studies, they attempted to standardize the methodology before making comparisons. Differences in assumptions included density of soil sampling, useful life of soil sampling information and cost of soil sampling, analysis, VRT application and recordkeeping. They found that the studies available at that time showed that VRT fertilizer was not profitable for low margin, extensive crops like wheat and barley, sometimes profitable for corn and quite profitable for higher value crops like sugar beet. They emphasized that the cost of soil nutrient information was a key factor in the results and that lowering the cost and increasing resolution of nutrient data with soil and optical sensors was a key requirement to widespread adoption of VRT fertilizer technology.

Subsequent, PA economic reviews by Lambert and Lowenberg-DeBoer (2000), Griffin et al. (2004) and Griffin and Lowenberg-DeBoer (2005) reinforced the conclusions of the Swinton and Lowenberg-DeBoer review. Lambert and Lowenberg-DeBoer reviewed 108 studies reporting economic analysis of PA technologies. Seventy-three percent of the studies reported on VRT technology mostly using partial budgets methods based on field trial data or simulation. Fewer than half (i.e. 44%) of the studies included information costs and only 30% included training and other human capacity expenses. Sixty-nine percent of the studies reported positive VRT fertilizer benefits. Four out of five sugar beet studies reported a benefit, 72% of the corn studies and only 20% of wheat studies.

Griffin et al. (2004) and Griffin and Lowenberg-DeBoer (2005) emphasized the economic advantage of 'embodied knowledge' technologies, compared with 'information-intensive' technologies like VRT. Most of the successful agricultural technology of the twentieth century was of the embodied knowledge type. For example, hybrid seed, fertilizer and most pesticides embodied substantial scientific knowledge, but did not require much expertise to use. Among PA technologies, GNSS guidance is an embodied knowledge technology and VRT fertilizer is a knowledge-intensive technology. They argued that VRT would need to become an embodied knowledge technology to achieve widespread adoption. One example of embodied knowledge VRT would be fertilizer application using a sensor mounted on the front of the machine, a computer using an algorithm on the applicator processing that sensor data, driving variable rate spreading out the back.

The economic studies in the first two decades of PA (starting from the first civilian use of GPS in the late 1980s) successfully identified the key economic opportunities for the technology, the long-term adoption patterns and the important constraints to widespread adoption. Those studies identified the potential of GNSS guidance and other locational technologies and questioned whether VRT as a knowledge-intensive technology could achieve general adoption. The economic studies were much less successful in identifying the short-term (i.e. year-to-year) patterns that speeded up or slowed down the adoption process in a specific geographic area, crop or farm type. The studies with a short term focus (e.g. Pierpaoli et al., 2013; Tey and Brindal, 2013) have been more successful in identifying the factors that could ex-post describe PA adoption trajectories. No one has been particularly successful predicting ex-ante short-run PA adoption patterns.

4 Predicting future trends based on recent studies of PA profitability

Understanding and predicting long-term PA adoption trends is important for farmers, agribusiness and research. Farmers need to make long-term plans when purchasing or leasing land, developing farm infrastructure and buying equipment. Understanding technology trends will help them make long-term decisions that facilitate the technology transition and avoid making major investments in technology that is about to become obsolete. Agribusiness profits often peak when supplying the technology and services demanded when large numbers of producers adopt a technology or innovation. Being ready to supply the innovation, and related inputs and services, is key to profitability. Research funding is often to be found for science that addresses key constraints limiting adoption of promising technologies. The PA economic studies of the last decade provide some insights on future PA adoption and are discussed in the following sections.

4.1 Sensor-based fertilizer VRT

Farmers intuitively like the idea of VRT fertilizer. It makes sense to manage different areas within fields according to the productive capacity and the constraints of that area. However, both economic analysis and the attempts to market VRT fertilizer services and equipment show that current technology does not realize the VRT vision for most producers of agronomic crops. The information costs needed to develop VRT fertilizer prescription maps is too high and the crop response may be limited by factors beyond soil nutrients. On-the-go sensor-based systems would make VRT fertilizer more like the embodied knowledge agricultural technology that has a track record of success. On-the-go sensor-based systems could augment accuracy by increasing the information resolution and reduce costs by turning the fertilizer prescription over to an algorithm.

Unfortunately, economic analysis of currently available on-the-go sensor systems suggests that improvement is still needed. For example, Biermacher et al. (2009) analysed on-farm winter wheat experiments in Oklahoma done during the period from 1998 to 2006. Those experiments tested an optical sensor and a nitrogen fertilizer optimization algorithm developed at Oklahoma State University. The study found that sensor-based nitrogen management has substantial economic potential, but the current system is about a breakeven with traditional nitrogen management. Boyer et al. (2011) analysed winter wheat trials using optical sensors to determine nitrogen fertilizer rates done in Oklahoma from 2005 to 2009. They found that yields with the optical sensors were often slightly lower than those in uniform rate treatments, suggesting that the sensor algorithm underestimated the nitrogen needed. Scharf et al. (2011) report that in fifty-five on-farm trials in the US state of Missouri, sensor-driven nitrogen application resulted in higher maize yields and lower nitrogen use than farmer's typical practices. Scharf et al. (2011) reported that on average, sensor-driven nitrogen application resulted in a US$42/ha increase in the value of the crop minus fertilizer cost, but did not deduct the cost of the sensor system and increased management complexity. An analysis of optical sensing-based nitrogen VRT for cotton production in three states in the United States showed that while the VRT reduced fertilizer use in some cases, it did not increase profits (Stefanini et al., 2015). Kuang et al. (2014) reported that an online visible and near-infrared spectroscopy of soil pH provided negligible economic benefits for barley production in Denmark.

One of the weak points of sensor-only systems for VRT fertilizer is that they do not take into account the productive capacity of the soil in adjusting fertilizer rates. For instance, crop colour may signal lack of nitrogen to the sensor, but in fact the problem is soil depth or another constraint. Diacono et al. (2013) reviewed 17 studies of PA for nitrogen management on wheat and recommended combining sensor data with soil maps, remote sensing images, yield maps and other data to tailor the nitrogen application to the site-specific constraints. They reported that, in the studies reviewed, on-the-go N sensing had a US$5–US$60/ha profit advantage over uniform application. McFadden et al. (2018) provide a Bayesian approach to including in-season information on yield potential in the VRT fertilizer algorithm. Some crop nitrogen sensor companies are starting to introduce systems that combine sensor data with other field information.

4.2 Agricultural robotics

Economic logic indicates that robotics should be adopted quickly when the technology is adequate for repetitive agricultural tasks like seeding, weeding and harvesting. Everywhere in the world (even in developing countries) farm labour is increasingly expensive and hard to find. Many developed countries (e.g. the United States, the United Kingdom) have depended on immigrant farm labour for many years, but governments face increasing political pressure to reduce both legal and undocumented immigration. Farms in developed countries are unlikely to be able to increase wages enough to attract local workers. Robots are the most likely alternative.

A key difference between the automation of conventional mechanized equipment which occurred with GNSS guidance, and agricultural robotics, is that a robotic agriculture is likely to use many relatively small machines instead of a few large ones (Auernhammer et al., 1995; Blackmore et al., 2004; Lowenberg-DeBoer, 2002; Gardener, 2011). Once machines are autonomous and the human beings are removed, the economic motivation for large machines almost disappears. Implementing cropping tasks with swarms of small robots will change agronomic practices and the geography of agriculture. For example with robotic pesticide application, it might be possible to spray each pest individually instead of broadcast application. This could reduce the amount of pesticide applied by over 90% and reduce the negative effects on beneficial species. With relatively small robots the economic advantage of large rectangular fields will be diminished. In the United States, the irregularly shaped fields with good soils, reliable rainfall and proximity to markets in New York and Pennsylvania could be competitive with the large rectangular fields of the Midwest and Great Plains that have more weather risk and greater distance to market. Small robots are less likely to trigger the safety and liability concerns that would keep humans on larger equipment to 'hit the off switch in case of a problem'.

Unfortunately, other than robotic milking, there is little on-farm data on robots in production agriculture. Milking robots were commercialized over 20 years ago. Numerous reports and studies have shown that they are practical and profitable for some dairy farms depending on labour cost, labour availability and other factors (e.g. Salfer et al., 2017; Hansen, 2015; Bijl et al. 2007; Armstrong and Daughety, 1997). For robots in crop production most economic studies have been based on economic engineering budgets, developed from technical parameters, not from field experience. For example, Pedersen et al. (2006) estimated benefits for weeding sugar beets, crop scouting in cereals and grass cutting on golf courses. They found that the robotic options were potentially economically competitive, with the cost of RTK-GNSS and the limited capacity of the small robots as the main constraints. With a similar methodology, Pedersen et al. (2017) found that early seeding with a small robot that could be in the field before soil conditions were dry enough for larger machinery and re-seeding with a similar robot would be profitable.

To put the robotics studies in context, from the earliest PA economic studies some observers have questioned profitability estimates based on simulation and economic engineering (e.g. Swinton and Lowenberg-DeBoer, 1998). The critics pointed out that simulation and economic engineering was a simplification of reality that would not reflect all of the constraints and risks faced on farms. However, Lambert and Lowenberg-DeBoer (2000) found that PA economic estimates based on simulation had roughly the same distribution of profitable/mixed/unprofitable results as those based on farm data.

4.3 Artificial intelligence and 'Big Data' in agriculture

AI and Big Data have been hot topics at agricultural research and industry conferences for several years. There are many academic papers on the subject. Coble et al. (2017) argued that analysis of Big Data will affect every stage in the agricultural value chain. Wolfert et al. (2017) found 114 refereed articles in English or Chinese relevant to the subject. An online search shows many blogs on the subject. In the United States agricultural input suppliers see data management as a growth area. In the 2017 CropLife survey, 58% of dealerships archived yield, soil and other data for farmers, 41% worked with farmers to analyse the data from their own farms, 17% pooled data across farmer customers and 10% participated in analysis of farm data beyond the dealership customer base. Only 31% said yield monitor data analysis was profitable for the dealership and only 20% indicated that profit/cost mapping was profitable. While it is logically appealing to use AI to make sense of the masses of data collected by sensors and other sources in agriculture, assessing the profitability and other economic benefits of this approach is more difficult. Problems in this economic assessment include:

- Defining exactly what constitutes Big Data – In most definitions, the key characteristic of Big Data is the massive amount of sensor data, but many mechanized farms have been collecting yield monitor data and accessing satellite images for years. Coble et al. (2017) argue that beyond size, Big Data in agriculture will be defined by volume, velocity, variety and veracity.
- What is AI in agriculture? – Computer algorithms have been used for many years in agriculture. For example since the 1960s, some farmers in the United States have used linear programming to plan crop rotations and make equipment choices (McCarl et al., 1977). Most animal feed in the industrialized world is formulated by computer algorithms.
- As with remote sensing, drones, yield monitoring and other information-gathering technologies in agriculture, the profitability of AI and Big Data analysis depend on the decisions made and actions taken. AI and Big Data analysis do not by themselves add to the bottom line.
- Most of the available economic analyses of AI and Big Data are based on simulation and economic engineering, rather than on field data because there are very few good examples of practical use of these concepts in commercial agriculture. For example, Li and Yost (2000) simulated the use of AI to optimize nitrogen and irrigation management, and showed a substantial economic gain. As in the case of agricultural robotics, the fact that much of the economic analysis is based on simulation rather than on-farm data is not enough of a reason to dismiss the idea, but it is a reason to press for more on-farm data.

4.4 Low-cost PA tools for medium and small farmers

So far, PA has been applied mostly on larger scale farms. One of the key challenges and opportunities is to adapt the technology for the millions of medium and small-scale farmers throughout the world (World Bank, 2017; Lowenberg-DeBoer, 2018). Boosting medium- and small-scale farm productivity through PA could help the world improve food security in the face of a population increase projected to more than nine billion. The science needed to adapt those PA tools for smaller, often unmechanized farms, is an opportunity for applied researchers. Providing the PA tools for those millions of farmers is a major opportunity for agribusiness.

The harbinger of the coming wave of PA tools for medium and small farmers is the adaptation of handheld optical sensors for use in managing nitrogen fertilizer in the developing world. Ortiz-Monasterio and Raun (2007) tested use of a sensor with wheat farmers in the Yaqui Valley in Mexico. The use of the sensor when combined with a nitrogen-rich strip for calibration saved farmers an average of US$56/ha over two crop seasons. Teboh et al. (2012) reviewed some attempts to use the N sensor technology in Africa. As with many innovations for agriculture in the developing world, one of the key challenges for PA is developing the supply chain that will provide technology, inputs and technical support at a cost that is within the purchasing power of medium and small farmers. The typical approach of many multinational agribusinesses is to sell a 'stripped down' version of their first-world technology in the developing world. Experience with other technologies suggests that going back to the original science and re-engineering a solution that fits developing countries conditions is often more successful than trying to sell first-world technology that was designed for very different conditions.

5 Future trends and conclusion

This chapter was built on the observation that widespread adoption of a technology is a good indicator of profitability or other economic benefits. With that point in mind, this chapter reviewed the publicly available data on use of precision agriculture in production of agronomic crops, identified the degree to which economic research has correctly identified PA technologies that would eventually be adopted by farmers and highlighted several PA technologies that recent economic research has identified as having profit, and consequently adoption, potential. The chapter focused on the economics of agronomic crops (e.g. cereals, oils seeds, cotton) because most PA economics research has focused on these crops, but the insights apply to PA for livestock, horticulture, forestry and other biological production systems. The chapter concentrated on PA economics for relatively large-scale mechanized farms because the economics of PA has been studied mostly for those types of farms, but has some observations about how the economics of PA might work out in medium and small farms.

PA adoption data show that GNSS guidance and related technologies like sprayer boom control and seeder row shutoffs are being adopted rapidly by mechanized farmers worldwide. Documented adoption rates are reaching the 70–80% range where a technology has become standard practice. Those same data show that use of VRT for fertilizer and other inputs is growing slowly. VRT is used by many producers of some specialized crops, like sugar beets in the US states of Minnesota and North Dakota, but for cereals, oilseeds and cotton it is difficult to find a region where more than 20% of farmers use VRT or 20% of cropland is managed with VRT.

A review of the economics research in the first two decades of PA suggests that by the late 1990s economists understood that GNSS guidance and related technologies were likely to be widely adopted in the longer run. By the late 1990s, economists had also started raising questions about the profitability of VRT for fertilizer and other inputs. They had identified the potential for reducing skip and overlap in input application to pay the GNSS guidance bills, and adoption trends to be reinforced by the reduced operator fatigue and other harder to quantify benefits. They had also raised questions about the cost of intensive soil sampling and other spatial information needed to make VRT decisions, the expense of human intervention in making VRT recommendation maps and the unpredictability of crop response to any of the inputs managed with VRT. PA economic research has been much less successful in understanding short-run adoption behaviour. A social science perspective has been able to ex-post identify some of the factors affecting PA adoption short run, but no one has been successful in predicting short-run PA adoption.

In the last decade economists have continued to study the economics of new PA technology as it emerged. The PA economics literature is now so voluminous that the kind of almost exhaustive study done by Lambert and Lowenberg-DeBoer (2000) and Griffin et al. (2004) is no longer possible. PA technology that shows economic promise and the potential for widespread adoption over the next decade includes sensor-driven VRT N, agricultural robotics, AI and Big Data, and PA for medium and small farms.

6 Where to look for further information

As the reference list for this chapter should show, there is a broad range of sources for PA economics. Some key sources of new developments are discussed in the following sections.

6.1 International Conference for Precision Agriculture (ICPA) and the European Conference for Precision Agriculture (ECPA)

There are numerous local and regional PA conferences, but the ICPA and ECPA usually provide the most comprehensive overview of PA research, including economic research. The proceedings of these conferences becomes a key record of the 'state of the art' in PA. The ICPA and ECPA are normally held in alternate years. Information about the next conferences can be found on the International Society of Precision Agriculture (ISPA) website: https://www.ispag.org/.

6.2 Refereed scientific journals

PA economics rarely captures the attention of academic economists and so it is rarely a topic in disciplinary economics journals. Advances in PA economics are more likely to be published in more applied subject matter journals like *Precision Agriculture*, https://link.springer.com/journal/11119; *Computers and Electronics in Agriculture*, https://www.journals.elsevier.com/computers-and-electronics-in-agriculture; or *Agricultural Systems*, https://www.journals.elsevier.com/agricultural-systems.

6.3 Agribusiness

Equipment manufacturers and software developers are eager to communicate benefits of their most recent products and services, but for a third-party view of the costs, benefits and potential profits, it is usually better to go to the farm media. It is their job to sift through the daily deluge of claims and identify those most worthy of attention. There are many to choose from worldwide, but in the United States they include:

- *Farm Journal AgTech* – https://www.farmjournalagtech.com/news
- *Farm Industry News* – http://www.farmindustrynews.com/
- *CropLife* magazine – focused on the agribusiness perspective – https://www.meistermedia.com/agribusiness-market/precision-farming/

7 References

Armstrong, D. V. and Daugherty, L. S. 1997. Milking robots in large dairy farms. *Computers and Electronics in Agriculture* 17(1): 123–8.

Auernhammer, H., Muhr, T. and Demmel, M. 1995. GPS and DGPS as a challenge for environmentally-friendly agriculture. *Journal of Navigation* 48(2): 268–78.

Awan, S. 2016. A review of the past, present and future of precision agriculture in the UK. Research Report No. 87. Kenilworth, UK: Agriculture and Horticulture Development Board (AHDB). https://cereals.ahdb.org.uk/media/1126475/rr87.pdf (Last accessed 1 May 2018).

Balafoutis, A. T., Beck, A., Fountas, S., Tsiropoulos, Z., Vangeyte, J., van der Wal, T., Soto-Embodas, I., Gómez-Barbero, M. and Pedersen, S. M. 2017. Smart farming technologies - Description, taxonomy and economic impact. In: Pedersen, S. and Lind, K. (Eds), *Precision Agriculture: Technology and Economic Perspectives. Progress in Precision Agriculture*. Dordrecht, the Netherlands: Springer.

Batte, M. and Ehsani, M. R. 2006. The economics of precision guidance with autoboom control for farmer-owned agricultural sprayers. *Computers and Electronics in Agriculture* 53: 28–44.

Bergtold, J. S., Raper, R. L. and Schwab, E. B. 2009. The economic benefit of improving the proximity of tillage and planting operations in cotton production with automatic steering. *Applied Engineering in Agriculture* 25(2): 133–43.

Biermacher, J., Brorsen, W. B., Epplin, F., Solie, J. and Raun, W. 2009. The economic potential of precision nitrogen application with wheat based on plant sensing. *Agricultural Economics* 40: 397–407.

Bijl, R., Kooistra, S. R. and Hogeveen, H. 2007. The profitability of automatic milking on Dutch dairy farms. *Journal of Dairy Science* 90: 239–48.

Blackmore, B. S., Fountas, S., Tang, L. and Have, H. 2004. System requirements for a small autonomous tractor. *Agricultural Engineering International: The CIGR Journal of Scientific Research and Development*. Manuscript PM 04 001. http://www.cigrjournal.org/index.php/Ejounral/article/view/525 (Last accessed 1 May 2018).

Boyer, C., Brorsen, B. W., Solie, J. B. and Raun, W. R. 2011. Profitability of variable rate nitrogen application in wheat production. *Precision Agriculture* 12: 473-87.

Castle, M., Lubben, B., Luck, J. and Mieno, T. 2017. Precision agriculture adoption and profitability. *Cornhusker Economics*. https://agecon.unl.edu/cornhusker-economics/2017/precision-agriculture-adoption-profitability (Last accessed 1 May 2018).

Coble, K., Mishra, A., Ferrell, S. L. and Griffin, T. W. 2017. Big data in agriculture: A challenge for the future. *Applied Economics Perspectives and Policy* 40(1): 79–96.

Cook, S. E., Adams, M. L. and Bramley, R. G. V. 2000. What is obstructing the wider adoption of precision agriculture technology? In: Robert, P. C., Rust, R.H and Larson, W. E. (Eds), *Proceedings of the 5th International Conference on Precision Agriculture*. Madison, WI: ASA-CSSA-SSSA (CDROM).

Department for Environment, Food and Rural Affairs (DEFRA). 2013. Farm practices survey autumn 2012 – England. https://www.gov.uk/government/uploads/system/uploads/attachment_data/file/181719/defra-stats-foodfarm-environ-fps-statsrelease-autumn2012edition-130328.pdf (Last accessed 1 May 2018).

Diacono, M., Rubino, P. and Montemurro, F. 2013. Precision nitrogen management of wheat: A Review. *Agronomy for Sustainable Development* 33: 219–41.

Erickson, B., Lowenberg-DeBoer, J. and Bradford, J. 2018. 2017 Precision agriculture dealership survey. West Lafayette, IN: Departments of Agricultural Economics and Agronomy, Purdue University. http://agribusiness.purdue.edu/precision-ag-survey (Last accessed 1 May 2018).

Franzen, D. W. 2017. Profitable use of site-specific nutrient management technologies. American Society of Agronomy. Annual Meeting Presentation 223-1. https://scisoc.confex.com/scisoc/2017am/videogateway.cgi/id/31003?recordingid=31003 (Last assessed 1 May 2018).

Gardener, D. 2011. Robotics and smart engineering. *Journal of Farm Management* 14(3): 249–70.

Griffin, T. and Lowenberg-DeBoer, J. 2005. Worldwide adoption and profitability of precision agriculture: Implications for Brazil. *Revista de Politica Agricola* 14(4): 20–38. https://seer.sede.embrapa.br/index.php/RPA/article/view/549/498 (Last accessed 1 May 2018).

Griffin, T., Lowenberg-DeBoer, J., Lambert, D., Peone, J., Payne, T. and Daberkow, S. 2004. Precision farming: Adoption, profitability and making better use of data. *2004 Triennial North Central Farm Management Conference*, Lexington, KY. https://www.agriculture.purdue.edu/ssmc/ (Last accessed 1 May 2018).

Griffin, T. W., Lambert, D. M. and Lowenberg-DeBoer, J. 2005. Economics of GPS lightbar navigation and auto-guidance technologies. In: Stafford, J. (Ed.), *Precision Agriculture-2005 Proceedings of the 5th European Conference on Precision Agriculture*. Wageningen, the Netherlands: Wageningen Academic Publishers, pp. 581–7.

Griffin, T. W., Shockley, J. M. and Mark, T. B. 2017a. Economics of precision farming. In: Shannon, D. K. and Clay, D. (Eds), *Precision Agriculture Basics*. Madison, WI: American Society of Agronomy, Crop Science Society of America and Soil Science Society of America.

Griffin, T. W., Miller, N. J., Bergtold, J., Shanoyan, A., Sharda, A. and Ciampitti, I. A. 2017b. Farm's sequence of adoption of information intensive precision agriculture technology. *Applied Engineering in Agriculture* 33: 1–7.

Hansen, B. G. 2015. Robotic milking - farmer experiences and adoption rate in Jaeren, Norway. *Journal of Rural Studies* 41: 109–17.

Kitchen, N. 2008. Emerging technologies for real-time and integrated agriculture decisions. *Computers and Electronics in Agriculture* 61: 1–3.

Kuang, B., Tekin, Y., Waine, T. and Mouazen, A. M. 2014. Variable rate lime application based on on-line visible and near infrared (vis-NIR) spectroscopy measurement of soil properties in a Danish field. *Proceedings of the AgEng Conference*, Zurich, Switzerland, 6–10 July 2014. http://www.geyseco.es/geystiona/adjs/comunicaciones/304/C03770001.pdf (Last accessed 30 April 2018).

Lambert, D. and Lowenberg-DeBoer, J. 2000. Precision farming profitability review. www.agriculture.purdue.edu/ssmc (Last accessed 1 May 2018).

Lawson, L. G., Pedersen, S. M., Sørensen, C. G., Pesonen, L., Fountas, S., Werner, A., Oudshoorn, F. W., Herold, L., Chatzinikos, T., Kirketerp, I. M. and Blackmore, S. 2011. A four nation survey of farm information management and advanced farming systems: A descriptive analysis of survey responses. *Computers and Electronics in Agriculture* 77: 7–20.

Li, M. and Yost, R. S. 2000. Management-oriented modeling: Optimizing Nitrogen management with artificial intelligence. *Agricultural Systems* 65: 1–27.

Llewellyn, R. and Ousman, J. 2014. Adoption of precision agriculture-related practices: Status, opportunities and the role of farm advisors. CSIRO. https://grdc.com.au/

resources-and-publications/all-publications/publications/2014/12/adoption-of-precision-agriculture-related-practices (Last accessed 1 May 2018).

Lowenberg-DeBoer, J. 1998. Adoption patterns for precision agriculture. Agricultural Machine Systems SP-1383. Warrendale, PA: Society of Automotive Engineers. Also available as SAE Technical paper 982041. https://www.sae.org/publications/technical-papers/content/982041/ (Last accessed 1 May 2018).

Lowenberg-DeBoer, J. 1999. GPS based guidance systems for agriculture. Purdue Agricultural Economics Report. Purdue University, December 1999, pp. 8–9. https://ag.purdue.edu/agecon/Documents/PAER_December%201999.pdf (Last accessed 1 May 2018).

Lowenberg-DeBoer, J. 2002. Liability problems for autonomous equipment? Site-Specific Management Center Newsletter (February 2002). www.agriculture.purdue.edu/ssmc (Last accessed 1 May 2018).

Lowenberg-DeBoer, J. 2018. The future of agriculture: Challenges and opportunities of precision agriculture for medium and small-scale farms. InterAmerican Development Bank. https://blogs.iadb.org/sostenibilidad/en/2018/01/10/the-future-of-agriculture/ (Last accessed 1 May 2018).

Lowenberg-DeBoer, J. and Swinton, S. 1997. Economics of site-specific management of agronomic crops. In: Pierce, F. J., Robert, P. C. and Sadler, J. D. (Eds), *The State of Site-Specific Management in Agriculture*. Madison, WI: ASA-CSSA-SSSA.

McCarl, B., Candler, W., Doster, D. H. and Robbins, P. 1977. Experiences with farmer oriented linear programing for crop planning. *Canadian Journal of Agricultural Economics* 25(1): 17–30.

McFadden, B. B., Brorsen, W. and Raun, W. 2018. Nitrogen fertilizer recommendations based on plant sensing and bayesian updating. *Precision Agriculture* 19: 79–92.

Méndez, A., Vélez, J., Villarroel, D. and Scaramuzza, F. 2014. Evolución de la Agricultura de Precisión en Argentina en los últimos 15 años (Evolution of Precision Agriculture in Argentina in the last 15 Years). *13th International Precision Agriculture Course*, 24–25 September. Manfredi, Argentina: INTA. http://www.agriculturadeprecision.org/13roCursoAgPrec/Libro/g4_Libro13CursoAP2014.pdf (Last accessed 1 May 2018).

Miller, N., Griffin, T., Bergtold, J., Sharda, A. and Ciampitti, I. 2017. Adoption of precision agriculture technology bundles on Kansas farms. Selected Paper. Mobile, AL: Southern Agricultural Economics Association, 4–7 February 2017.

Mitchell, S., Weersink, A. and Erickson, B. 2017. Precision agriculture in Ontario: 2017 precision agriculture services dealership survey. Working Paper Series - WP 17-01. Canada: Institute for the Advanced Study of Food and Agricultural Policy, Department of Food, Agriculture and Resource Economics, University of Guelph.

Molin, J. 2016. Precision agriculture in Latin America. *INFO AG Conference*. https://www.infoag.org/presentations/2378.pdf.

National Research Council. 1997. *Precision Agriculture in the 21st Century: Geospatial and Information Technologies in Crop Management*. Washington DC: National Academy Press.

OECD. 2016. Farm management practices to foster green growth. OECD Publishing. http://dx.doi.org/10.1787/9789264238657-en

Ortiz, B. V., Balkcom, K. B., Duzy, L., van Santen, E. and Hartzog, D. L. 2013. Evaluation of agronomic and economic benefits of using RTK-GPS-based auto-steer guidance systems for peanut digging operations. *Precision Agriculture* 14: 357–75.

Ortiz-Monasterio, J. I. and Raun, W. 2007. Reduced nitrogen and improved farm income for irrigated spring wheat in the Yaqui Valley, Mexico, using sensor based nitrogen management. *Journal of Agricultural Science* 145: 215–22. doi:10.1017/S0021859607006699.

Pedersen, S. M., Fountas, S., Have, H. and Blackmore, B. S. 2006. Agricultural robots - System analysis and economic feasibility. *Precision Agriculture* 7: 295–308.

Pedersen, S. M., Fountas, S., Sørensen, C. G., Van Evert, F. K. and Blackmore, B. S. 2017. Chapter 8: Robotic seeding: Economic perspectives. In: Pedersen, S. M. and Lind, K. M. (Eds), *Precision Agriculture: Technology and Economic Perspectives*. Heidelberg, Germany: Springer International Publishing.

Pickthall, T. and Trivett, E. 2017. An investigation into the barriers that prevent the adoption of precision farming technologies in combinable cropping in the UK. *Aspects of Applied Biology 135: Precision Systems in Agricultural and Horticultural Production*, pp. 29–37.

Pierpaoli, E., Carli, G., Pignatti, E. and Canavari, M. 2013. Drivers of precision agriculture technologies adoption: A literature review. *Procedia Technology* 8: 61–9.

Robert, P., Rust, R. and Larson, W. 1994. Site-specific management for agricultural systems. In: *Proceedings of the 2nd International Conference on Precision Agriculture*. Madison, WI: ASA-CSSA-SSSA.

Salfer, J., Endres, M., Lazarus, W., Minegishi, K. and Berning, B. 2017. Dairy robotic milking systems - What are the economics? University of Minnesota Extension. http://articles.extension.org/pages/73995/dairy-robotic-milking-systems-what-are-the-economics (Last accessed 1 May 2018).

Scaramuzza, F., Villarroel, D. and Vélez, J. P. 2017. Incremento de las herramientas de agricultura de precisión. Mayor automatismo y precision para la sustentabilidad del ambiente productivo (Increase in Precision Agriculture Tools. Greater Automation and Precision for the Sustainability of the Productive Environment). *16th International Course in Precision Agriculture and Livestock*, 29 September 2017. Manfredi, Argentina: INTA E.E.A. http://www.agriculturadeprecision.org/16toCursoAgPrec/Libro/03_Scaramuzza_F_et_al_Incremento-de-las-Herramientas-de-Agricultura-de-Precision.pdf (Last accessed 1 May 2018).

Scharf, P., Shannon, D. K., Palm, H. L., Sudduth, K. A., Drummond, S. T., Kitchen, N. R., Mueller, L. J., Hubbard, V. C. and Oliveira, L. F. 2011. Sensor-based nitrogen applications out-performed producers-chosen rates for corn in on-farm demonstrations. *Agronomy Journal* 103(6): 1683–91. http://extension.missouri.edu/sare/documents/ProducerStudies2012.pdf (Last accessed 1 May 2018).

Schimmelpfennig, D. 2016. Farm profits and the adoption of precision agriculture. Economic Research Report 217. USDA Economic Research Service.

Stefanini, M., Larson, J., Boyer, C., Cho, S-H., Lambert, D. and Yin, X. 2015. Profitability of variable rate technology in cotton production. Selected Paper. *Southern Agricultural Economics Association Annual Meeting*, Atlanta, GA. https://ageconsearch.umn.edu/bitstream/196995/2/SAEA_CottonVRT.pdf (Last accessed 1 May 2018).

Swinton, S. M. and Lowenberg-DeBoer, J. 1998. Evaluating the profitability of site-specific farming. *Journal of Production Agriculture* 11: 439–46.

Teboh, J., Tubaña, B. Udeigwe, T., Emenback, Y. Y. and Lofton, J. 2012. Applicability of ground-based remote sensors for crop N management in Sub Saharan Africa. *Journal of Agricultural Science* 4: 175–88. http://www.ccsenet.org/journal/index.php/jas/article/view/11757 (Last accessed 1 May 2018).

Tenkorang, F. and Lowenberg-DeBoer, J. 2008. On-farm profitability of remote sensing in agriculture. *Journal of Terrestrial Observation* 1(1): 50–9. https://docs.lib.purdue.edu/jto/vol1/iss1/art6/ (Last accessed 1 May 2018).

Tey, Y. S. and Brindal, M. 2013. Factors influencing the adoption of precision agricultural technologies: A review for policy implications. *Precision Agriculture* 13: 713–30.

Thorp, K. 2014. Precision agriculture. In: Njoku, E. G. (Ed.), *Encyclopedia of Remote Sensing. Encyclopedia of Earth Sciences Series*. New York, NY: Springer.

USDA. 2018. Tailored reports: Crop production practices. https://data.ers.usda.gov/reports.aspx?ID=17883 (Last accessed 1 May 2018).

Watson, M. and Lowenberg-DeBoer, J. 2004. Who will benefit from GPS auto guidance in the Corn Belt? Purdue Agricultural Economics Report. http://www.agecon.purdue.edu/extension/pubs/paer/2004/paer0204.pdf (Last accessed 1 May 2018).

Whelan, B. M. and McBratney, A. B. 2000. The 'Null Hypothesis' of precision agriculture management. *Precision Agriculture* 2(3): 265–79.

Wolfert, S., Ge, L., Verdouw, C. and Bogaardt, M.-J. 2017. Big data in smart farming – A review. *Agricultural Systems* 153:69–80.

World Bank. 2017. *ICT in Agriculture (Updated Edition): Connecting Smallholders to Knowledge, Networks, and Institutions*. Washington DC: World Bank. https://openknowledge.worldbank.org/handle/10986/27526 (Last accessed on 1 May 2018).

Zarco-Tejada, P., Hubbard, N. and Loudjani, P. 2014. Precision agriculture: An opportunity for EU farmers – Potential support with the CAP 2014-2020. Directorate General for Internal Policies, Policy Department B: Structural and Cohesion Policies, European Parliament Committee on Agriculture and Rural Development. http://www.europarl.europa.eu/RegData/etudes/note/join/2014/529049/IPOL-AGRI_NT%282014%29529049_EN.pdf (Last accessed on 1 May 2018).

Zhang, N., Wang, M. and Wang, N. 2002. Precision agriculture – A worldwide overview. *Computers and Electronics in Agriculture* 36: 113–32.

www.ingramcontent.com/pod-product-compliance
Ingram Content Group UK Ltd.
Pitfield, Milton Keynes, MK11 3LW, UK
UKHW021843100926
8556IPUK00019B/1027

9 781801 466028